PROSPERITY FROM TECHNOLOGY

A New Approach to Industrial Production, Money and the Environment

David Rudd

The Book Guild Ltd
Sussex, England

The Book Guild Ltd.
25 High Street,
Lewes, Sussex

First published 1999

Set in Times
Typesetting by
Acorn Bookwork, Salisbury, Wiltshire

Printed in Great Britain by
Antony Rowe Ltd, Chippenham, Wiltshire

A catalogue record for this book is
available from the British Library

ISBN 1 85776 362 9

PROSPERITY FROM TECHNOLOGY

CONTENTS

ACKNOWLEDGEMENTS

My thanks are due to three friends for reading the chapters, more or less as I wrote them, and offering their comments and criticisms. They are Frank Edmead, who had a distinguished career as a reporter and leader writer on a national broadsheet, Arne Johansson, my Swedish opposite number on a research project in Brussels in 1983–5, and Don Soughan, a technologist friend since we were both apprentices. Needless to say they are not responsible for the book's shortcomings, but without their encouragement I should not have finished it.

I have also to thank Pauline Barley for the cartoons and, last but not least, my wife for all her help and forbearance throughout the whole endeavour.

1

INTRODUCTION AND PREVIEW

This book is written by a technologist, mainly for non-technologists, but its purpose is not just to explain the mysteries of technology to the lay person. Its purpose is to derive from an understanding of modern technology some important features which a society needs in order to prosper from it.

Every society throughout history and prehistory has depended on its ability to make practical use of its knowledge, however acquired. The early prehistoric societies discovered how to make fire and tools, how to catch animals, when and how to prepare the ground and plant seeds and so on. They could not have survived without that ability, but they acquired it piecemeal – often by chance. They lacked the necessary prior knowledge on which to base the systematic acquisition of further knowledge. At the other end of the scale, the technology of modern industrial societies is derived from modern science, which is a highly systematic body of knowledge. That systemisation has enabled modern science to make colossal strides in the last 100 years, and indeed in the last 25 and the last five years.

The importance of technology in a society which aspires to high living standards is not likely to be disputed in the so-called Western world. The benefits in North America, Western Europe and many other places are plain for all to see, but it is easy to lose sight of the extent to which technology pervades the lives of all the members of the modern industrial societies, rich and poor alike. The anonymous but famous man in the Clapham omnibus may be wishing he could afford a car and a skiing or wind-surfing holiday in a sunnier clime but, even if he is unemployed, he is far better fed, more comfortably clad and more pleasantly housed than his forebears would have believed possible. He takes for granted both the smoothness of the road on which he is travelling and the television programme he watched the night before. Evidently those societies possess many of the necessary features for prosperity. But it is sadly true that those benefits have come about inconsistently and capriciously and have brought, or seemed to bring, great misery to some people, even within those comparatively prosperous societies. And of course other societies have conspicuously failed to derive much benefit from modern technology; some are worse off than they were before they came into contact with it.

As an extreme example of such a failure, our hearts are wrung today by the plight of millions of starving people in Africa, but they have not reached that condition by remaining completely out of touch with modern technology or deliberately excluding it from their way of life. Had they done so, they would probably now have the same rather short life expectancy as their forebears and suffer from the depredations of wild animals and some diseases which modern medicine (which is highly technological) has eliminated or reduced, but at least there would be far fewer of them and they would probably not be starving. It must be said that the mistake in those societies was to try to apply modern technology without understanding how to do so. No doubt in some cases the misery has been caused or exacerbated by the malevolence of the ruling factions in those societies or their indifference to the plight of their subjects, but even when their intention – or the intention of the outsiders introducing the technology – has been beneficent, they have too often failed from their lack of that understanding.

Between the extremes of horrible poverty and almost indecent prosperity, one can find every gradation of success and failure in producing and distributing the necessities, comforts and luxuries of life, and every kind of unwanted consequence in the way of malnutrition, poor housing, unemployment, disaffection and damage to the environment.

Together they constitute a vast field of controversy in which politicians, economists, commercial entrepreneurs, journalists, social scientists, philosophers and latterly environmentalists freely roam, but technologists are inclined to wait until they are consulted as specialists. Not unexpectedly, these non-technologists have started from their own experience and brought in the necessity, desirability or iniquity of technology as and when they perceived its relevance; they have not set out to discover those features which are specific to the derivation of prosperity from technology, irrespective of other objectives.

For the most part, at least in Britain, the leading speakers and writers in all those categories have at best a superficial acquaintance with technology of any kind and so do not perceive that there might be such features. They lack the knowledge from which they might look for that possibility. On the other hand most technologists (numerically speaking) take little really serious, sustained professional interest in anything outside their own technology. They may worry rather vaguely about its undesirable effects 'out of office hours' at their institution meetings and so on, but that is as far as they get. Yet it is arguable that the way to avoid these undesired consequences, enhance the degree of prosperity and spread it more widely is to be found through understanding how technologists go about their business – the information they use and how they use it, particularly their sometimes unconscious assumptions and habits of mind.

The emphasis is on *through* such understanding. It will not be contended that the understanding will be enough by itself, but that it will provide a new angle of vision and a new point of departure from which those problems can be attacked. That is the approach this book adopts.

'Prosperous', like many common words, is incapable of precise definition[1], but nowadays it means much more in the industrialised countries than it did even 100 years ago, when anyone who did not have to worry about where his or her food, clothing and shelter were coming from was regarded as prosperous, at least by the great majority, who did have to worry. Perhaps an acceptable definition today would be something like 'having a good deal of what you want, or what you think you want, in the way of goods, services, leisure and pleasant surroundings, now and in the foreseeable future'. The starting point of this book is that modern technology has increased our material living standards so much that prosperity in that sense has become an aspiration of all who do not already have it, whereas, historically, not very long ago it was accepted to be the prerogative of a privileged few. And that does not imply prosperity at the expense of other members of a society or of other societies, but by the creation of wealth.

This definition of prosperity does not imply that prosperity necessarily brings happiness, nor that such things as family cohesion, social stability and national defence are less important than prosperity. But material prosperity and how to achieve it or increase it are major preoccupations of most of us for much of our time, probably because we believe or assume it will bring the other things in its train, or at least that it will facilitate their acquisition and retention. There are many books about the validity or otherwise of that belief. This one will concentrate for the most part on those features which are necessary or desirable for prosperity, leaving aside the features which may be necessary for other purposes, except for a short comment in the penultimate chapter.

* * *

Four hundred years ago, Francis Bacon could write: 'I have taken all knowledge to be my province'[2], but that was 50 years before the birth of Isaac Newton. The subsequent explosion of knowledge has eradicated the possibility of such a claim being taken seriously today, and we are left in the situation that our biggest problems are approached with an inadequate individual understanding. That is the dilemma from which we must begin. C. P. Snow (later Lord Snow), the famous novelist and scientist, came near to its heart 40 years ago in his lecture entitled 'The Two Cultures and the Scientific Revolution'[3]. His theme was that our Western literary intellectual culture has become detached from modern science, and that consequently scientists have developed a separate culture – he called it 'the scientific culture' – which does not recognise the superiority of the literary intellectual culture. He contended that the inability of the two groups to understand each other is an enormous loss to our society.

Snow certainly understood and emphasised that prosperity has become an aspiration (in his view a legitimate aspiration) of all who do not already have it, and many technologists would concur with his theme and would recognise that they are much more imbued with the scientific culture than with the literary intellectual culture. And he understood the difference between science and technology, but he was not concerned to follow it up. The difference arises from the absence of purpose in pure science, or – which amounts to the same thing – that its purpose is to accumulate scientific knowledge without regard to its utility. Pure scientists assert that their discoveries and theories are not to be judged by their relevance to the needs or wishes of the societies in which they are made or formulated. *Scientia gratia scientiae* (knowledge for the sake of knowing) might be their motto. Very well, but that is where technology parts company with pure science. Technology (or applied science in Snow's terminology) is a separate discipline, in which the essential principle is the application of knowledge for a stated purpose. And the consequences of the difference between pure science and technology are far-reaching.

The purpose of technology is not always to create prosperity. Sometimes it is or has been to win a war, or to put a man on the moon, or to glorify God, for example by building cathedrals, which were great technological achievements, and there are no doubt other purposes to which it can be put. But we shall be concerned only with the purpose of contributing to the prosperity of the society in or by which the technology is employed. That purpose generates a series of interactions between technology and its adjacent disciplines – accounting, commerce and economics, politics, environmental studies and philosophy, and it is those interactions which are the main concern of this book.

That does not imply that a person can be classified simply as a technologist or a physical scientist, accountant, economist, entrepreneur, politician, environmentalist or philosopher. Perhaps some people can be so restrictively classified, but most people pick up knowledge and wisdom from other disciplines and occupations than their own and they are also influenced by their private and social friendships and contacts. Some people have two or three quite different occupations in their working lives and large co-operative endeavours increasingly involve multi-disciplinary teams. But the cores of the disciplines have become and remain largely separate and distinct. Our goal in this book will be to understand some important effects of the disciplines on one another and to tease out and resolve some conflicts between their very roots.

* * *

When a technologist ventures outside his[4] primary field into an already well-populated (some would say overcrowded) territory, he must expect to be challenged by the prior occupants to show his credentials and state his business. His reply is that in the course of their normal work he and his colleagues are taking millions of decisions, some large, many small, which singly or cumulatively have enormous and often long-lasting consequences. He will add that their decisions, or most of them, are rational and that they are even adept at the difficult business of planning under conditions of uncertainty; indeed their rationale is a large, intricate, intellectual edifice. But he is worried about two things.

If he is allowed to continue, the technologist will explain that his first worry is that he has long suspected the quality of the information which he and his colleagues are given or otherwise acquire as the raw material of their decisions; and his second worry is that the foregoing intellectual edifice has been constructed gradually on a foundation of assumptions (technologists are prone to such metaphors) about how society in general, and the economy in particular, operates and he is not sure that the foundation is sound. Most of his colleagues do not worry on either score, probably because technology itself is so fascinating that they have become addicted to it and are now 'technoholics'.

The technologist will go on to say that after carefully examining the edifice, he is convinced that it cannot be reconstructed to accommodate these deficiencies. Of course the challenger is welcome to come inside and learn enough about its construction to verify that for himself – and to propose modifications, which will be carefully considered if they do not prevent the edifice from fulfilling its main function. But unless some remedial action is taken, the harmful effects of technology will increase uncontrollably and may begin to outweigh the benefits, even in the comparatively prosperous societies (they have already done so in some poorer societies). In this situation he has some proposals for improving the quality of the information which goes into the edifice and for modifying and strengthening its foundations to support its ever-increasing weight.

But suppose that the technologist's first statement – that his and his colleagues' decisions have serious consequences – is gainsaid by the challenger, who asserts that the big decisions, which have such repercussions, are taken by senior managers, economists and government ministers who are, by and large, not technologists. In that event, the technologist might be stung to retort that that is just the trouble; they ought to be technologists. But such an answer would be unwise because, although it may be true, it is not the whole truth. The additional point is that, no matter who takes the supposedly big decisions, they are actually no more than choices between a few options prepared by technologists in more lowly positions, and all the options are affected by the quality of the technologists' information and the reliability of their assumptions. It is of course necessary for sceptics to read this book to understand why that is so.

* * *

The search for the specific features which a society needs in order to prosper from technology is naturally bound to encounter many problems, but there is the additional complication that the main problems are linked together. They must therefore be tackled together to enable the solutions to any of them to be perceived and their relevance to the others explained, but they are not all in the province of any one traditional profession or occupation. So the technologist outside his primary field is not just wandering about at random, praising, criticising and innovating as he fancies and collecting friends, foes and sceptics as he goes. He has a steady purpose, which is to bring the threads of the arguments together in the end, but the order of the chapters in this book has been chosen so as not to entangle those threads more than can be helped on the way. A few words in advance about the construction of the book and the chapters may help to explain what is going on.

It does not seem possible to begin without implying that the economy must have a monetary base. Barter economies cannot prosper from modern technology. (Probably they cannot prosper, in the sense in which we are using the term, without modern technology, but that is another question.) But some space will be devoted to analysing *why* technologists must use money – its special significance for them, which is additional to the significance it has for non-technologists. That analysis will uncover the route to other conclusions. It will consider for example how wide the monetary base must be and the vexed question of whether any parts of the economy can use technology and contribute to the general prosperity without being based on money.

The link between technology and economics will probably be readily admitted, although it is stronger than is generally appreciated, but the link with accounting may occasion some surprise, not least among accountants. The short explanation is that every science must have its 'metrology', by which is meant its procedures for deciding what to admit as data and how to judge their accuracy and reliability. For physical data, technologists can rely on Metrology with a capital M, the highly developed science of measuring lengths, masses, times and other physical quantities, which serves the science of physics and serves it extremely well.

But Metrology provides only half the data which technologists must have. The other half – their costs, prices and money values – must come from elsewhere since physics has no need of such measurements and Metrology is silent about them. The obvious source is accounting because accountants keep track of money transactions, which are the only primary source of all measurements and estimates of costs, prices and money values. Unfortunately annual accounting, which is the dominant branch of accounting[5], has been developed without much regard for the specific needs of technology, and it is flawed in consequence.

Accounts are also the only primary source of data for economics, and annual accounting has paid no more attention in its development to economics than to technology. But by and large economists seem content – or perhaps resigned – to accept statistical and other secondary and tertiary sources of data, without delving as deeply as perhaps they might into all the steps between the raw data of actual money transactions and the information which surfaces in their theories and recommendations.

Technologists do understand their own need for a second metrology, though they do not habitually express the thought in those words, and they have partly overcome the lack of relevance of annual accounting by setting up a separate metrology, which exists alongside accounting, often in the same firm, and gets much of its raw data from the invoices which pass through the technologists' hands for verification on their way to the accountants for payment. We shall see that the technologists' metrology is a developed science, in some respects more so than annual accounting, and that, although it is partly complementary to accounting, it operates in territory which is also within the province of accounting. But the technologists' metrology does not embrace all the relevant transactions and their stratagem has left the principal flaws in annual accounting untouched.

Just how present-day annual accounting practices frustrate the aims and undermine the benefits of technology and the reforms which are needed are the subjects of Chapters Four and Five. But a necessary precursor must be to explain how technologists come to their decisions and the assumptions which they make, so those explanations are given in Chapters Two and Three. The aim in Chapter Two is to impart the way technologists in the prosperous societies think rather than what they think about, but thought processes are difficult to describe in the abstract. It is much easier, for the reader as well as the author, to consider how technologists deal with a particular set of problems and to extract their thought processes and mental habits in the process.

To carry conviction, the problems must be in a fairly advanced but not too obscure field of technology and an obvious candidate is public electricity supply. In the industrialised countries everybody uses electricity at home and probably at work as well, often in large quantities; grid lines and towers are familiar sights and generating stations are in evidence in most parts of Britain and other industrial countries. These are good qualifications for the short list of candidates. The additional qualification which gives electricity supply the edge over other contenders is that it can also be used to illustrate the links between technology, accounting and commerce. Because electrical energy cannot be stored[6], electricity supply poses special commercial problems, whose peculiar difficulty has been a contentious subject for the best part of a century.

Chapter Three is about how technologists measure and estimate their costs, using their second metrology, and how they evaluate new projects and allow for monetary inflation. Annual accounting then occupies two chapters because there are two large problems which accountants have still not solved, even to their own satisfaction, although, as a body, they appear to be more easily satisfied than the users of accounts. One is the notorious problem of accounting for monetary inflation, which is not of course the same thing as controlling or preventing inflation. The other is how to apportion the capital cost of an item of plant or equipment among the successive years of its service life – the problem of annual depreciation. It does not impinge very seriously on how technologists measure and control their costs, and they have not given it much attention, but it is serious in the management of a company or commercial undertaking which depends on technology and in the determination of its prices and methods of charging, especially if it is publicly owned or regulated. Moreover one aspect of it is important in inflation accounting. Therefore Chapter Four covers annual depreciation in the absence of inflation and Chapter Five deals with accounting for inflation.

If the thoughtful reader is ready to grant that physicists presumably need Metrology and that technologists need reliable data about costs, he or she may nevertheless suspect that the latter subject is going to be very dull. Accounting has that reputation. If so his or her interest may be quickened when it is pointed out that the tenets of Metrology are not so obvious as perhaps they first appear. Lengths, masses and times cannot always be measured as easily as they are in a factory or a shop. If you happen to be travelling very fast or accelerating very rapidly (and it may not be immediately obvious whether you are doing so or not), your measurements will be grossly affected. That is part of what Einstein announced in his Theory of Relativity, after he had given a great deal of attention to an examination of how physicists went about their business – the information they used and how they used it, particularly their (until then) often unconscious assumptions and habits of mind. Until he did so, Metrology was probably a dull subject, but Relativity transformed physics and led eventually to nuclear fission, nuclear bombs and nuclear power, which have been described in many ways but are certainly not dull subjects. One should perhaps hasten to say that the necessary reforms to accounting are not to be compared in complexity with the Theory of Relativity; and concomitantly it will not be asserted that they will necessarily lead to anything as revolutionary as nuclear fission, merely that the eventual ramifications of a thorough re-examination of first principles are sometimes more extensive – and therefore more interesting – than is realised when the re-examination is begun.

Chapter Six is mainly about the commercial aspects of public electricity supply in public ownership, as an example of the interactions between technology and commerce, particularly the requirements which technology imposes on commercial undertakings. It draws on the conclusions of Chapters Two through Five to derive solutions to a commercial problem which is commonly ignored or glossed over or sometimes stated to be insoluble. The solution depends on the theory of annual depreciation formulated in Chapter Four and the method of accounting for inflation in Chapter Five, which are therefore essential to obtaining the greatest prosperity from commercial applications of technology.

Chapter Seven is in two parts. The first part analyses the problem of regulating natural or technical monopolies when they are in private ownership, and the weakness of the only two extant methods of regulating their prices and profits. A formula is proposed to induce the shareholders and work force not only to economise but also to share the benefits of the economies with the customers, with the minimum of external supervision.

The second part of Chapter Seven goes on to deal with the regulation of an oligopoly, defined as a group of a few companies providing a common product or service where there are not enough companies in the group to bring about genuine competition between them. It challenges the view that competition is the only effective method of regulating the prices charged by the companies in an oligopoly and proposes instead an extension of the formula for regulating monopolies to apply also to oligopolies.

Chapter Eight, 'Technology and Politics', invokes the principles set out in the earlier chapters, first to show why communism will always fail to bring prosperity from technology, even in a society which is initially benign, and then to comment on public versus private ownership of important services in market economies and the treatment of physically limited resources in a market economy. Chapter Eight also covers how technology affects non-commercial and quasi-commercial public services, the vexed question of remuneration in a free society which relies on technology, with a sub-section on trade unions, and the link between technology and unemployment.

One of the ills for which technology is regularly blamed is damage to the environment. On a superficial or restricted conception of prosperity, its pursuit often does damage the environment. For example, if a community which lives by a lake unthinkingly excludes the condition of the lake from its efforts to prosper from modern technology, the lake will sooner or later become heavily polluted by sewage and industrial effluent and the real prosperity of the community will suffer correspondingly, but it is imperceptive to blame technology for that effect, although in some circumstances you might legitimately blame the technologists. Obviously the community should include the present and foreseeable future condition of the lake, or in general society should include the present and foreseeable future condition of the environment, in its conception of prosperity. Technology can then become the means of preserving and perhaps improving the environment instead of destroying it. The approximate definition of prosperity earlier in this present chapter included the phrase 'in pleasant surroundings, now and in the foreseeable future' with that consideration in mind.

Some industrial societies have already taken that first vital step, but the next steps are much more difficult. Must the features which are derived in the earlier chapters, particularly the emphasis on costs, be dismantled in the interest of preserving our planet Earth in a habitable state, and shall we then have to submit to arbitrary decrees laid down by environmentalists? Technologists have no panacea for protecting the environment but Chapter Nine describes some progress towards resolving that apparent conflict and points to the way forward.

For some readers, particularly in the already prosperous societies, Chapter Nine may provide the main – or even the sole – justification for this book. If one already has plenty to eat and drink, a large and well-furnished house, a wardrobe full of nice clothes, ample means of private transport, good medical services, good schools for one's children and enjoyable holidays, why should one worry about more prosperity? Many people have turned, and many more will turn, their attention away from increasing their own or anyone else's material prosperity towards more congenial pursuits – the arts, sport, exploration, etc. – long before they reach that condition. The answer does not depend on altruism or philosophical introspection. It is that some recent phenomena now threaten the prosperity of us all, both rich and poor, or at least the prosperity of our children or grandchildren. Three are selected to illustrate the theme. In order of decreasing parochiality and increasing global importance, they are urban air pollution, acid rain and global warming. It is one thing to disdain to examine how one's prosperity has been produced and might be increased when one's material necessities, comforts and luxuries seem to be secure; it is quite another thing to continue in that frame of mind when there is a real prospect that they will be seriously undermined in the foreseeable future.

Chapter Ten, entitled 'Technology and Philosophy', explores the links between those two subjects in the light of the earlier chapters. One of the lessons of modern philosophy seems to be that fundamental essays in that subject are inclined to run eventually into the sand. But if so, it will be contended that the sand is not so deep as to obliterate some residual but necessary links between philosophy and technology which are often not mentioned – and may therefore be forgotten – in contexts where they are literally vitally important. Chapter Ten also reverts briefly to the question, which was side-stepped at the outset, about prosperity, happiness, family cohesion, social stability, defence and so on.

Chapter Eleven brings the threads of the earlier chapters together and summarises the conclusions.

* * *

In a book which is intended to bring together branches of knowledge which are normally regarded as separate, the individual chapters must not assume any prior expert knowledge of the subjects with which they are severally concerned, but they are not intended to give just the flavours of the subjects. Their aim is to explain the subjects in sufficient detail to readers who may be experts in one or more of the other subjects, and to go on from there to the assumptions and practices which are hampering the processes of sustainable wealth creation, the links between them and their counterparts in the previous chapters, and the proposed reforms. That approach has the additional advantage that the arguments can then also be understood by persistent general readers with no expert knowledge of any of the subjects. The understanding of those general readers may be a necessary step on the way to implementing those reforms, if – as unfortunately seems likely – the established experts do not respond to the proposals for reform or fail to perceive the links between their own and other branches of knowledge.

As a consequence of that approach, some readers will inevitably come across passages which seem obvious, even banal, but the question to be asked then is whether those passages will seem equally obvious to other readers from different occupations and backgrounds. Such is the variety of knowledge and beliefs which people have and hold that they probably will not, in which case, if the passages are relevant to later conclusions, there is no way of reaching those conclusions except via those passages. Such readers are therefore requested to skim-read when they come to obvious passages, but to bear in mind that they are necessary parts of the arguments. There is no point in nodding through a passage but then ignoring the conclusion(s) to which it leads.

It must also happen that some conclusions will have been reached by writers and others in the past without special reference to technology, but have been disputed by opposing schools of thought. In these cases the intention is to provide additional reasons, not generally advanced, for supporting one side in the controversy rather than the other. Finally, as far as the author is aware, many of the arguments are original. They have mostly been previously published by the author but in limited contexts. Here the opportunity is taken to explain their relevance as links between technology and its adjacent disciplines. This book is thus partly a guided tour of territories well known to their occupants and partly a series of traverses across some comparatively unmapped borderlands.

It is, of course, open to any reader to take the chapters in any order, and to break off in the middle of a chapter because its relevance to their concerns is not apparent. Some may even be willing to accept the conclusions of a chapter without bothering with the detailed arguments. To assist both categories, if they then come across assertions in later chapters which seem illogical or incredible, there are back-references to the passages in earlier chapters from which those assertions have been derived.

At this point, if not before, some readers may start to wonder why the author writes about 'technologists' instead of using the shorter and more familiar word 'engineers'. The reason is that many words have two meanings, a common meaning and a special meaning, and a wise speaker or writer avoids the special meaning if he or she is addressing a non-specialist audience or readership. For example, I am a Bachelor of Science, but I am married and I did not read Science (which is not the same as Technology) at university; also I am a Doctor of Philosophy, but I do not treat sick people and I have not taught Philosophy. I am a Chartered Engineer, a Fellow of the Institution of Electrical Engineers and a Member of the Institution of Mechanical Engineers, but I do not agree with most of my fellow members that we should refer to ourselves as engineers or Engineers outside our own counsels. My reason is that we do not drive railway trains, repair motor cars, washing machines or TV sets or otherwise work most of our time with our hands, which is what engineers in the ordinary sense of the word do. If I went round calling myself a bachelor or a doctor, I should only create confusion and be told, quite rightly, not to be silly. It seems to me as silly to call myself an engineer. I am not, of course, alone in this view; a number of organisations, though not yet those two institutions, have stopped calling technologists engineers.

Part of the training of a technologist involves manual work and the acquisition of some manual engineering skills, for example how to join pipes together by flanging and bolting or by welding. Those skills are relevant to the technologists' choices, when fully qualified, between different methods of construction and operation, just as the skills of bricklaying and concrete mixing are (or should be) relevant to architects' choices between those two methods of building. But there is much more to those choices than the levels of manual skill they require and it is as confusing to call technologists engineers (and vice versa) as it would be to call architects bricklayers.

Modern technology makes extensive and sometimes intensive use of mathematics, though not as intensive as physics. No technologist can expect to get far in his or her profession without a grounding in the subject, but such knowledge is not necessary in many walks of life and it is not essential for understanding the arguments in this book, if the reader is prepared to take some things on trust. There are, therefore, no mathematical formulac in thc main text.

However mathematics is a language and mathematical arguments which are presented in non-mathematical terms are like translations from a foreign language into the native language of the reader. If you read a good translation of *The Brothers Karamazov*, you may hope for a passable understanding of Dostoyevski's feelings, and if you also read translations of half a dozen other Russian authors, carefully selected, you may begin to understand the history which led up to the October revolution, but you will depend on the translator's skill for that understanding. Your insight would be deeper and more authoritative if your education had enabled you to read the original works in the Russian language. Similarly a knowledge of mathematics would deepen the reader's understanding of those arguments in this book.

Fortunately many people nowadays who are neither technologists nor scientists have an adequate – often ample – knowledge of the not very advanced mathematics employed in the passages where such formulae have been invoked in this book. To cater for them the formulae have been put into appendices and footnotes, along with a miscellany of technical and other supplementary information and explanation, which would be more distracting than helpful if it were included in the main text.

It is also not essential for an understanding of the arguments in this book to have any liking for graphs or diagrams, but people to whom they are anathema are probably by now in a rather small minority. For the majority, who find graphs helpful, there are some in Chapters Four, Six and Seven. They cannot be put into footnotes but they can be ignored without any great loss of continuity by those who positively dislike them – provided, of course, that they can follow the arguments without them.

It is not one of the aims of this book to instruct the reader in the jargon of technology, or indeed any of the jargons of accounting, commerce, economics, politics, environmental studies or philosophy. The intention is to use words in their ordinary sense, except where refinements are essential to carry the argument further than ordinary language can penetrate, and then to introduce and define as few special terms as possible. However the jargon is often entrenched and has spread outside the disciplines in which it originated. In those cases it is mentioned for the convenience of readers who are already familiar with it, and sometimes adopted, albeit with reluctance.

REFERENCES AND FOOTNOTES

1) The *Chambers English Dictionary* definition is 'thriving, successful', which is fair but not very helpful because the definition of 'thrive' in that dictionary is 'get on, do well, prosper'.
2) In a letter to Lord Burleigh in 1592.
3) 'The Rede Lecture', 1959, *Cambridge University Press*.
4) In this instance the technologist is the (male) author.
5) The other branch is cost accounting, generally known as or included in management accounting. This criticism is not directed against management accounting, *per se*, although the links between management accounting and technology are regrettably weak.
6) Except in infinitesimal quantities in electrical capacitors. It can be converted into chemical energy and stored in batteries or into potential energy and stored in high-level water reservoirs, but the conversions and reconversions back into electrical energy lose some energy and the batteries and converters or reservoirs, pumps, turbines and ancillary equipment are comparable in cost with the generating stations which produce the electrical energy in the first place.

2

THE NATURE OF TECHNOLOGICAL DECISIONS

Anyone who has not studied physics much since they left school may assume, perhaps unconsciously, that the design and construction of an electricity generating station are entirely governed by physical laws. They probably remember that one can make electricity flow round an electrical circuit by spinning a magnet inside a coil of wire to which the circuit is connected, and that the energy which is required to spin the magnet, or most of it, reappears in whatever apparatus is connected into the circuit. They may even remember that the process was discovered by Michael Faraday in 1831 and that the physical law is one of Faraday's Laws of Electromagnetic Induction.

Such an assumption is quite consonant with knowing or learning that, in the electricity supply industry in Britain and most other industrial countries, the source of the energy is usually coal, oil, natural gas or nuclear fuel[1] and that in most existing stations the conversion into electrical energy is an elaborate business, involving burning the fuel under a boiler (or inducing the corresponding nuclear reaction) to raise high-pressure steam and then jetting the steam on to the blades of a turbine, which drives the spinning magnet[2]. Such stations are known as 'thermal' stations and the physical laws governing these processes are so numerous that they may appear at first sight to determine the materials and dimensions of the apparatus and probably the buildings in which it is housed as well, so that any serious departure from the specifications which these laws require would cause the station to fall down or the apparatus to explode or injure the operators in some way – or at any rate fail to produce electricity in the intended quantity. Alternatively, since electricity generation is an energy conversion process, it may seem that the aim must be to make the conversion as efficiently as possible, 'efficiency' being defined as the ratio of the electrical energy sent out from the station to the heat energy of the fuel which it consumes.

THE SCOPE FOR VARIATION IN DESIGN

But these appearances are misleading. It is true that physical laws govern the processes and therefore bear on the design of the apparatus and the buildings, and unfortunately there are sometimes more or less dramatic failures of apparatus or buildings, which the technologists must strive to prevent. Also efficiency is an important factor which is constantly in the forefront of the design process. But that is not the whole story, not even most of the story, because although physical laws govern the processes they do not by any means specify the design of the apparatus or the buildings, nor is efficiency the paramount factor. Within the limits of what will stand up and produce enough electricity with an acceptably high degree of safety there is great scope for variation, including substantial departures from maximum efficiency, and it is with the designers' choices within those limits that we shall mainly be concerned.

A station which fails to produce enough electricity is an obvious sign of incompetence, or inadequate technology. The same is true of a serious accident, which may be the subject of a public enquiry. But the contribution of a station to the prosperity of the community in which it operates is not so obvious. It is quite possible for the construction and operation of a safe station over a 25-year operating lifetime to make little or no such contribution. It may leave the community poorer than if it had been made much less efficient, and even sometimes poorer than if it had not been built. And we shall see that in some circumstances there may be very little that the technologists, in designing the station, can do about it.

The word 'design' may evoke comparisons with designing furniture or clothes, and it may seem surprising that the same word can be used for such different activities without confusion, bearing in mind the care that has been taken in Chapter One not to use the same word, 'engineer', to describe people who apply technology and people who drive trains or repair cars, TV sets or washing machines. But there is a real sense in which designing generating stations is like designing, say, women's clothes, in fact more like designing women's clothes than like driving trains or repairing cars, TV sets or washing machines.

Of course it would be absurd to suppose that any clever designer of women's clothes could design a generating station, or vice versa. But both are constrained by physical laws, which determine, for example, the strengths and elasticities of materials, their resistance to heat and water and how they can be joined together, and both have great scope for choosing within those constraints; also both are essentially planners of work for others to do. We should not seriously offend normal parlance if we defined 'to design' as 'to choose how to make [something] within the relevant physical constraints'. People design clothes, furniture, motor cars and generating stations in that sense. The essential difference between them is how they exercise those choices.

Within the limits of their materials and methods of tailoring, designers of women's clothes presumably strive for elegance, chic (if that is not the same thing), distinctiveness, fashion, beauty perhaps, and attractiveness to the opposite sex no doubt. People may say that they buy clothes to keep themselves warm and dry, but in a prosperous society they willingly buy far more clothes than they need for those two purposes. It is their demands for quantity, variety, attractiveness and so on which keep the designers busy. (The same is true of men's clothes, but not to quite the same extent, even today, though the gap is narrowing.)

Generating station designers sometimes speak of elegance – even beauty – and their sentiments are sometimes shared by non-technologists, but the designers' main objective is quite different. A good generating station *produces* – or contributes to the production of – prosperity, whereas good clothes worn by people in a prosperous society are a *manifestation* of their prosperity. We might classify generating stations as 'producer goods' and clothes as 'consumer goods', but that would be a poor classification because it is far from obvious which class we should put furniture, motor cars, aeroplanes, bicycles or word processors into. All, or nearly all, products have producer and consumer aspects in varying proportions. It is more illuminating to put generating stations and women's clothes at or near the opposite ends of a scale, along which the position of any good or service depends on the importance which is attached to pure utility in its design and production or provision.

'Utility' is a useful word up to a point, but we shall have to come eventually to the hard questions: how do the designers of generating stations convert a vague conception of utility into specifications of materials and dimensions of plant and equipment (the preferred terms for 'apparatus') and buildings, and how ought they to do so? We might answer the first question straight away by saying that (within the limits of safety and security of supply and with due attention to quality) they try to minimise the cost, but unfortunately 'cost' has two distinct meanings, both legitimate. Sometimes it means what the buyer or user pays or foregoes for something, which sounds simple but actually requires some thought, and at other times it means what the community foregoes in order to have that something, which sounds – and is – much less simple but can hardly be ignored if we are looking for prosperity. One example is the cost of petrol (gasoline). In Britain three quarters of what the motorist pays is tax, which is a so-called transfer payment from the motorist to the Treasury, not requiring the community as a whole to forego anything in order to supply the motorist; only the other quarter represents what the community must forego. Both kinds of cost have their staunch advocates in many applications of technology with billions of pounds (or dollars, or whatever) at stake, but there is nothing to be gained and much to be lost in plunging into that controversy before we have studied the plant and equipment a little more closely than we did at the beginning of this chapter.

We have a second reason for postponement here. There are strong and influential bodies of opinion which view the use of cost as a criterion in important decisions with grave suspicion. In Chapters Eight and Nine we shall enquire as to whether there are any other criteria which could be substituted for cost in some applications of technology and, if so, what they are, when and how they might be applied and whether they conflict with the aim of increasing our prosperity. But first we need an open-minded understanding of the kinds of decisions technologists must make and the physical constraints within which they must make them. It will be illuminating to see how far we can get by common sense, without bringing in any costs.

A SIMPLE CHOICE?

We can begin with a simple – or apparently simple – choice. What shall we use to conduct the electricity in the generators[3] and between the generators and the consumers? Knowing that we are dealing with an energy conversion process, we shall keep efficiency in the forefront of our minds as a matter of common sense. ('Efficiency' can have different meanings in different contexts, but in this chapter it will mean efficiency of energy conversion in the sense already defined.)

At the present time only a few purified metals have suitable conducting properties. That is the main physical constraint within which the designer must make his or her decision. All the properties of those metals, including their strength, elasticity, ductility and chemical properties, come into the calculations eventually, but we can make a good start by looking at their electrical resistance and weight. In a generator the weight is not crucial but the heat produced by the current flowing against the resistance must be removed, and it represents a loss in the total energy conversion, so that the boilers and turbines have to be a little larger, and considerably more fuel has to be burned than if the conductors had no resistance.[4]

The conductors in a generator are rectangular in section rather than round and they have holes down the middle for cooling water or oil, but the relevant dimensions are their length and their cross-sectional area (excluding the holes). For a typical conductor with a length of 6 metres and a cross-sectional area of 16 square centimetres, we can compare the resistance, the power loss when carrying a typical current of 15,000 amps and the weight of each candidate metal on our short list of purified metals. They are:

Metal	*Resistance (micro-ohms[6])*	*Approximate power loss[5] (kilowatts)*	*Weight (kilogrammes)*
Silver	56	12.7	100
Copper	59	13.2	86
Gold	77	17.2	185
Aluminium	92	20.7	26
Iron	334	75.0	76

Silver heads the list and we could certainly use silver if we based our decision solely on the physical properties of the metals, ignoring the consequences for the rest of the community, but we have no authority nor any wish to do that. Let us imagine that we are in a small community with a limited population and natural resources, including some coal.

From the above figures we can calculate that to make 50 conductors (for a typical generator) we should need 5 tonnes[7] of silver or 4.3 tonnes of copper, but because of their higher resistance the copper conductors would necessitate the removal of 4% more heat and the station would burn a little more coal. However, suppose that in exploring our domain we find a mountain with a large deposit of copper/silver ore containing 0.7% of copper and 0.014% of silver[8]. It follows that for 50 silver conductors the miners in our community would have to dig up 36,000 tonnes of ore, but only 600 tonnes for 50 copper conductors.

The difference between 36,000 tonnes and 600 tonnes persuades us to choose copper. Although we realise that the extra coal will probably amount to two or three thousand tonnes over the life of the station, common sense will persuade us that our community will be more prosperous if we use copper than if we use silver, even if we are not sure whether any good use can be found for the 80 kilograms of silver which will also be produced. Thus we begin to relinquish the aim of maximum efficiency, but as yet only slightly.

There are two important points which this first example brings out without our referring to any costs:

i) the advantage of the decision to the community is realised not at the generating station or in the designer's office but at the copper/silver mine, which may be hundreds of kilometres away;
ii) the form of the advantage is almost entirely a large reduction in the number of ore miners employed, or in the time taken to produce the conductors with a given number of miners, or probably some combination of both; the disadvantage, which common sense has told us will be smaller, is a small increase in the number of coal miners required (at a third location) throughout the service life of the station.

In reality, the chain of links between the decision and its consequences is usually longer and more obscure than in this example, being part of an elaborate network of chains between myriads of analogous decisions on other questions affecting that station and all the other applications of technology and their equally remote consequences. But it is essential that the chain shall not be broken, for if it is broken the decision will not contribute to the prosperity of the community. To take an extreme case, if the ore miners were going to dig up 36,000 tons of ore, regardless of the demand for either metal, perhaps because they would otherwise be idle and that would be regarded as bad for their morale, it would be better to make the conductors of silver. That would at least save some coal, which would be welcome – unless of course the coal mine were operated with the same disregard for the level of demand for its product, but that would hardly contribute to the prosperity of the community.

A curious but true story may help to dispel a possible impression that the foregoing example is a put-up job and that the choice of the conductor material is really determined by the laws of physics. It concerns a group of technologists who were beleaguered and had to construct not a generator in this case but some very large electro magnets[9]. They could not get hold of nearly enough copper, but they discovered a 'mine' of pure silver, not silver ore, which they used successfully for the conductors in their electro-magnets.

It happened in the Second World War when the Allies were trying to make the first nuclear fission bombs. The fissionable substance for one design of bomb was uranium, but uranium has two isotopes (elements which have the same chemical properties but different nuclear properties), only one of which is fissionable. A kilogramme of natural uranium contains only 7 grammes of the fissionable isotope, U235; the remaining 993 grammes are the other isotope, U238, and its presence inhibits the fission of the U235 isotope. So the bomb depended on separating the two isotopes, but they could not be separated chemically because they have the same chemical properties. Two separation methods were tried, one of which involved shooting a stream of electrically charged atoms of natural uranium into a powerful magnetic field. The field deflected the slightly lighter atoms of U235 a little more than the slightly heavier U238 atoms, so the atoms could be made to fall into two separate boxes, one for the U235 atoms and the other for the U238 atoms. A set of enormous electro-magnets was constructed to create the magnetic field and they had to be wound with thick conductors carrying large currents. In wartime there was a severe shortage of copper, but there was enough silver in the vaults of the US Federal Bank and it was lent out for that special purpose. That was technology for war, not prosperity, and that separation method proved to be less effective than the other method, but not because the silver conductors were inferior to copper conductors. In fact they were slightly superior by virtue of their lower resistance.

Looking further afield, we find that silver is comparatively scarce in the earth's crust[10], in fact so scarce that using it habitually to make conductors for generators would have stifled every electricity supply industry in the world at birth, and we should all be much poorer than we are. Gold is, of course, even scarcer and it has no redeeming physical properties for our particular purpose, while iron, though plentiful and lighter than copper, has a much higher resistance. So why not just say that scarcity is the deciding factor within the other physical constraints, instead of dragging in networks of chains of decisions and hinting about unemployment among ore miners?

No doubt many technologists would like to distance themselves from any hint that their decisions may lead to unemployment, but the scarcity argument will not wash. Aluminium is much more plentiful in the earth's crust than copper or any of the other candidates, yet technologists do not use aluminium for the conductors of large generators and, in the present state of technology, we should not be more prosperous if they did. The objection to aluminium is partly that it has a higher resistance than copper but mainly that the chemical difference between aluminium and its ore creates much more serious problems than the difference between copper and its ore.

In a generator, we might consider compensating for the higher resistance of aluminium conductors by making the conductors wider and/or thicker. For the same resistance they would have to be 60% larger in cross-sectional area than copper conductors. That would make them still only half as heavy as copper conductors, but their greater bulk would necessitate making the whole generator much bigger, with more iron to provide paths for the magnetic fields, which must surround the conductors, and much more steel to hold the whole thing together, so it would be much heavier and would need correspondingly stronger foundations and a larger building. On the other hand, for the conductors between the generators and the consumers bulk is not important and the lighter weight is a big advantage, especially for the overhead lines because the towers have to support the weight of the conductors, so lighter conductors mean fewer, less sturdy towers.

Turning to the chemical problem, aluminium is or was much more difficult to extract from its ore than copper or silver. So much so that in the time of Napoleon III aluminium metal was scarcer than gold and the emperor is reputed to have possessed a dinner service with aluminium plates, which were held to give greater distinction to his banquets than the gold plates at the banquets of other rulers[11]. The production of aluminium metal on a commercial scale did not begin until the latter part of the nineteenth century, whereas copper has been smelted since prehistoric times. In the first half of the twentieth century, aluminium was used for non-electrical purposes, including dinner plates (which were much used by campers but were deemed to be inferior to gold or china plates for banquets, presumably because they had become very cheap), and for overhead transmission lines, where the advantage over copper is greatest.

In Britain aluminium began to be used for underground and indoor cables after the Second World War, but there was a setback due to an unexpected corrosion problem; then, when that problem had been overcome, the jointers, trained in the craft of soldering copper, had to re-learn their trade; industrial appliances had to be redesigned to take the new cables; lack of foresight and precise instructions (easy to condemn in retrospect) sometimes delayed the completion of large capital projects and gobbled up the expected benefits. Some technologists lost a lot of sleep and promotion; a few may have lost their jobs; but for the most part the new cables were judged to be a success. How should we decide whether that judgement was sound?

Before we begin, there is a further complication. Copper and silver are often, though not invariably, found together, and they can then be extracted from the ore and separated in one fairly simple process. So in discussing the choice between them, the mining and extraction processes could be lumped together without seriously distorting the argument. To be more exact, we should have said: 'for 50 silver conductors the miners in our community would have to mine 36,000 tonnes of ore *and the smelters would have to extract the metal from that ore*, but only 600 tonnes for 50 copper conductors', but the stark contrast between 36,000 and 600 tonnes was not exaggerated by the omission. Aluminium, however, is not found in the same ore as copper or silver; moreover the modern commercial extraction process for aluminium consumes a great deal of electricity, which the smelting of copper and silver does not require. So the choices between copper and aluminium have been and still are inevitably bound up with the rate of growth of electricity supply systems.

A situation in which that choice could realistically be portrayed in terms of the physical properties which we now perceive to be relevant, namely the resistance and weight of the metals and the scarcity and chemical properties of their ores, without reference to any non-physical quantities such as price or cost, would have to contain much more than just some miners and smelters. It would have to include at least some generating stations and electricity grids and some separate plants for extracting the metals, with their respective technologists and all the grades of skilled and unskilled workers required for their design, construction and operation. Even if we ignored the complications of international trading and confined the illustration to one country, it would be too complicated and hedged about with qualifications for us to judge from the portrait whether it would be better to use copper or aluminium for any particular application.

It seems that by trying to keep an open mind and judge all the consequences, we find ourselves so encumbered with things to be borne in our open mind that we cannot take one of the simplest decisions in designing an electricity supply system. We are reduced to guessing, but if we guess badly we shall not prosper from our decision as well as societies which are able to make a rational judgement. We might try to copy them and some societies do just that, but the conditions are never the same in different places and they change with time, so the decisions in those societies are never as good as they might be.

And we have only just begun. We have to choose all the other materials for the generators, the boilers, the fuel and ash handling plant, all the ancillary plant and equipment, the foundations, the buildings and the distribution network – and then the dimensions and methods of manufacture and construction of all those items – hundreds if not thousands of different materials and methods and millions of dimensions. If we want the station to contribute to our prosperity we must base all our decisions on their consequences in those (mostly remote) parts of the economy which will be affected by them.

But perhaps the decision has been selected for its exceptional underlying complexity, concealed by an apparent simplicity, and perhaps the other decisions, though numerous, are much more dominated by the laws of physics or the aim of maximum efficiency. It is probably not possible to satisfy any deeply suspicious reader on that sort of point; some things have to be taken on trust; but the next and last example, which is taken from a different part of the generating station, may go some way to doing so.

USING HEAT TO SAVE HEAT

Let us therefore move now into the middle of the station and look at a rather ingenious feature of the turbine, which takes steam from the boiler and uses it to drive the spinning magnet in the generator. It is not a particularly modern feature but it is particularly relevant in the present context because the great majority of thermal generating stations – even the latest natural gas-fuelled stations – have steam turbines which offer the designers the opportunity of incorporating this feature.

In a fairly modern, coal-fuelled generating station[12], the steam goes into the turbine at a pressure of some 160 times the pressure of the atmosphere and a temperature of about 565°C, and it emerges at a tiny fraction of the pressure of the atmosphere and a temperature in the region of 35°C, at which point it is condensed into water. When this generator is producing 660,000 kilowatts of electricity, a tonne of steam, with a volume of 50,000 cubic metres, goes into the condenser underneath its turbine every three seconds or so. It comes out as water, with a volume of one cubic metre, and is pumped back into the boiler as feedwater to go round the boiler-turbine-condenser cycle again. But it is not pumped straight back into the boiler because, by putting some additional equipment between the condenser and the boiler, the designer can reduce the quantity of heat required by the turbine, and hence the quantity of fuel required by the boiler.

We have nearly reached the zenith – or nadir if you prefer – of technical complexity and it may help to get over it (or past it) if we realise that the paramount properties of the steam in a turbine are not its pressure but the temperatures at which it goes in and comes out. There is a physical principle which says that these must be as far apart as possible for maximum efficiency[13] and that we must remove a large quantity of heat, which inevitably escapes from the conversion process, at the outlet temperature. So the inlet temperature is as high as our materials can withstand and the outlet temperature is as little as we can conveniently make it above the temperature of the surrounding atmosphere, into which the surplus heat is discharged. We have to put up with the enormous volume of the outlet steam but we can reduce the volume of the inlet steam to manageable proportions by raising the pressure of the water before we boil it.

Now comes the ingenious part. The boiling temperature of water depends on its pressure. At one atmosphere it is 100°C but at 160 atmospheres it is 345°C and we have somehow to heat the feedwater from the 35°C at which it comes out of the condenser up to 345°C before we can boil it. We could, of course, design the boiler to do that, but there is another physical principle which says in effect that, when you are converting heat energy into mechanical energy, it is more efficient not to use heat at a high temperature if there is a convenient source of heat at a lower temperature which will do the job[14]. And there is a convenient source, in fact a number of convenient sources, in the turbine itself.

A turbine has some resemblance to a windmill – or rather a long line of windmills one behind the other and with steam blowing through them instead of air. In more formal language, the steam expands through the turbine in a series of steps, each comprising a circle of nozzles from which jets of steam issue and impinge on a row of blades on the rim of a spinning wheel, after which the steam is collected up at a lower pressure and temperature and put through the next circle of nozzles and on to the next row of blades, and so on. The key words are 'at a lower temperature'. We can think of the steps in the turbine as a temperature staircase, down which the steam falls from the boiler to the condenser and gives up some of its heat energy to a row of spinning blades at each step. But we are not obliged to put the steam in at the top of the staircase and wait for it all to fall out at the bottom. We can tap into the turbine at any step, extract some of the steam and use it to heat up the feedwater on its way back to the boiler.

The extracted steam is called 'bled' steam. In giving up its heat to the feedwater, the bled steam condenses into water and joins the main feedwater stream. We lose the mechanical energy which that bled steam would have produced if we had left it in the turbine to fall down the rest of the staircase, but we gain because the boiler is now fed with hotter water and so burns less fuel. The physical law says that the gain is bound to be greater than the loss. If that sounds to you like a conjuring trick, you are no different from many students when they first hear about bled-steam boiler-feedwater heating, but thousands of generating stations are very noisy witnesses to the absence of any trickery.

Of course the bled steam will not heat the feedwater right up to its own temperature; there must always be a temperature difference to induce the heat in the steam to flow into the feedwater. If we bleed the turbine at one point, part way down the temperature staircase, that will heat up the feedwater to about 150°C and increase the efficiency by about 6.5%. With two bleed points, the feedwater temperature will be raised in two stages, from 35°C to about 110°C in the first stage and from 110°C to 180°C in the second stage, with a further gain in the efficiency of about 2.5%. With eight bleed points, spaced down the staircase, the efficiency will be about 12.5% better than with none. Each stage requires a feedwater heater, rather like a mini-condenser, and some pipes, valves, ancillary equipment and insulation, plus of course some space in the building, and it all has to be kept in good order and repair. Also the high-pressure end of the turbine has to be larger, but the low-pressure end, the condenser and the boiler can be smaller.

So we have to decide how many stages of feedwater heating to instal. For technical reasons, which need not concern us, we really must have at least one and we cannot have more than the number of steps in the temperature staircase (in this turbine 26) but between those limits we have a free choice.

Shall we say 12 stages, or shall we enlarge the building and put in 26 stages, or shall we go for simplicity and limit ourselves to two or three?

If you were considering a gadget for saving petrol in your car, you might judge that a saving of anything less than 5% was not worth bothering with (many motorists would say 10%), but then one car does not consume much fuel. A typical motorist in Britain, who does an annual mileage of 15,000km (10,000 miles) uses only about 1,200 litres in a year, so 5% would be 60 litres, which is less than 45 kilogrammes. If the motorist sells the car after five years, his or her total saving will be something over 200 kilogrammes with luck, but everyone knows these gadgets are unreliable. On the other hand the boiler for a 660,000 kilowatt generator will burn 1.5 million tonnes of coal or oil a year, so 5% would be 75,000 tonnes and even 0.1% would be 1,500 tonnes a year. And feedwater heaters are very robust and reliable; also the service life of a generating station is seldom less than 25 years and often much longer. We shall not prosper if we design our generating stations with the same rough and ready rules of thumb which many of us use (not unreasonably) for private motoring.

The actual choice of British designers in 1986 was eight stages. The efficiency could have been slightly improved by increasing that number and there are dozens of other ways by which it could also be improved. For example, a larger condenser or a larger cooling tower (which discharges the surplus heat into the atmosphere), or both, would improve it by lowering the outlet temperature of the steam, so increasing the height of the temperature staircase. But what would the community actually gain? What would it lose if there were fewer feedwater heaters, a smaller condenser and smaller cooling towers?

THREE MISCONCEPTIONS ABOUT ENERGY

The trouble with people is not that they don't know but that they know so much that ain't so.[15]

Unfortunately, since the oil crisis in 1973, numerous careless, misleading, sometimes false, statements about energy and the supposed virtues of high efficiency in all parts of the economy have been bandied about in the newspapers (not just the popular ones), on television, in parliament, and even in some government departments and reputable technical journals in Britain and other countries. These statements have led to confusion and to three misconceptions, which have to be dispelled before we can assess the true importance of efficiency in judging the contribution of a generating station to our prosperity.

Energy is measured in joules,[16] or more conveniently in megajoules, which are millions of joules. Generally, applications of technology consume energy and when there are two methods of achieving any particular purpose, the quantities used by the different methods can be compared. The first misconception is that, because both quantities are expressed in the same units, megajoules, those comparisons must be valid comparisons of like with like.

For example, in Britain the transport sector of the economy is a large consumer of energy and much effort has been devoted to reducing that consumption. To illustrate the misconception, one might estimate that a typical small delivery van with a petrol (gasoline) engine will consume 100 megajoules of energy on a representative urban mission, whereas a similar van propelled by an electric motor supplied from a battery would consume only 35 megajoules on that mission, i.e. along the same route in the same time with the same payload. But that comparison has no relevance to any practical problem because energy, like matter, has different forms. The petrol van will require its 100 megajoules in the form of nearly 2.9 litres of petrol, whereas the electric van would require its 35 megajoules in the form of 9.7 kilowatt-hours of electricity.

Thirty-five – or even 100 – megajoules of energy in the form of petrol would be of no more use to an electric van than 35 or 100 kilogrammes of chalk to a chef who is waiting in the kitchen to prepare cheese soufflés for a banquet. And by the same token a supply of electrical energy would be of no more use to a petrol van than a parcel of cheese to a schoolteacher who wanted to write on the blackboard. Yet just such estimates were requested in Parliamentary Questions after the 1973 oil crisis, with the evident intention of urging the superiority of electrical propulsion for urban transport. What other intention could the questioners have had?

Perhaps not all the questioners were labouring under that particular misapprehension. Perhaps some of them had in mind that petrol can be turned into electricity – or more strictly a proportion of the energy which is, in a sense, 'in' petrol can be converted into electrical energy – whereas no-one knows how to convert chalk into cheese. It is true that the analogy between forms of energy and forms of matter breaks down in that respect. No analogy is perfect. But there is a second, subtler misconception in the way that possibility of conversion is habitually presented.

The normal presentation introduces the rather woolly, though often useful, concept of 'primary energy' – that is energy which is obtained from deposits in the earth's crust or from mountain lakes, rivers, the sun, the wind, the tides or sometimes hot subterranean rocks. The misconception is that, although it is misleading to compare quantities of different kinds of energy directly, it is legitimate and relevant to practical problems to compare them indirectly in terms of the primary energy which they supposedly represent. Many technologists who would scathingly point out the first misconception have unfortunately failed to point out the second, which is consequently far more widespread.

Coal and petroleum[17] are sources of primary energy and they can both be converted into electricity, with almost equal facility, in generating stations with steam turbines as described at the start of this chapter. In practice a refinery product known as 'residual oil' is normally preferred to petroleum but that does not affect the argument. Technologists, who really should know better, have been persuaded from time to time to say how many megajoules of electrical energy can be made from 100 megajoules of residual oil. In other words they have given a single figure for the efficiency of generation.

In 1977 one such figure was 32%[18]. Now, 100 megajoules of residual oil and 100 megajoules of petrol represent slightly different quantities, 103 and 115 megajoules respectively, of primary petroleum energy. So, allowing 9% for losses in the grid, the 35 megajoules for the electric van would seem to represent about $\{(35+9\%) \div 32\} \times 103 = 123$ megajoules of primary energy, whereas the 100 megajoules for the petrol van represent 115 megajoules of primary energy. *Ergo*, provided the mission is truly representative (which we shall not query), it would seem that electric vans would consume slightly more primary energy than petrol vans, but not much more. There have been scores, if not hundreds, of such estimates since then (though few as well documented and therefore as easy to criticise), many of them finding that electrically propelled vehicles of one kind or another consume, or would consume, less primary energy than petrol or diesel vehicles on the same missions. However, the reports containing estimates of either tendency seldom, if ever, say or ask why the efficiency was 32%, or whatever figure the estimators came up with.

We have seen in the previous section that the efficiency of generation is not determined by any physical laws and that it is not as high as the designers know how to make it. The efficiency in the cited report was the average of the efficiencies of a number of stations, whose individual efficiencies were just as high or as low as their designers chose to make them. Their choices varied from less than 10% to about 36%, but the bulk of the electricity came from the more efficient stations so the average was nearer the upper end of the range[19]. The same applies, to some extent, to the designers of the refineries, which convert the petroleum into the various refinery products, and the designers of the grid, which transmits the electricity from the generating stations to the battery of the van when it is being recharged. They can also choose how efficient to make their processes, but their choices are much more restricted – and so less relevant to the argument – than the choices of the generating station designers.

The laws of physics allow a range of options in the design of generating stations, but to claim that their efficiencies were 'just as high or low as the designers chose to make them' may sound like a gross exaggeration in the light of the decisions we have considered so far – the materials of the conductors, the number of boiler feedwater heaters and (in passing) the sizes of the condensers and the cooling towers. Those decisions can affect the overall efficiency of the station by only a few percentage points either way. However that was because we peeped into the middle of the design process instead of starting from the beginning. Let us quickly retrace our steps.

The boiler/steam-turbine/generator method is not the only method of generating electricity from petroleum. Another established method is to burn kerosene (another petroleum product) and jet the hot gases produced by the combustion directly on to the blades of a gas turbine, which also has a temperature staircase like a steam turbine but which does not, of course, have any boiler, condenser, cooling tower or boiler feedwater heaters. That is an inherently simpler but less efficient method than the steam-turbine method and the designers can tilt the simplicity/efficiency balance down to a very low efficiency if they wish to achieve the simplest possible design.

In fact they tilted it down to less than 10% for the simplest gas-turbine station and up to nearly 36% for the most elaborate and efficient steam-turbine station. And they could have widened the range if they had chosen to do so, for example by constructing combined heat and power systems, which use some of the heat rejected in the conversion process for space heating in housing estates and commercial premises, thereby raising the efficiency (measured by the ratio of useful energy output to energy input) to 80% or so.

So the assertion about the primary energy consumption of electric vans was really no more than a consequence of how technologists choose to design generating stations. That is not to say that their decisions are whimsical – quite the contrary – but we shall not prosper very much from technology if we choose electric vans for urban deliveries because they seem to save energy or reject them because they do not, while the generating stations, which will supply our vans, continue to be designed on what is evidently a quite different – and possibly incompatible – criterion.

There is yet a third method of generating electricity, albeit from coal more easily than petroleum, which relies on a completely different set of physical principles. In stations with steam turbines and/or gas turbines, the fuel is first burned to produce heat, then the turbines convert the heat into so-called 'mechanical' energy (the energy which spins the magnets in the generators) and finally the generators convert the mechanical energy into electrical energy. But why make three conversions if one will do? In a device known as a 'fuel cell', the energy of a refined coal product, methanol, can be converted directly into electrical energy with an efficiency of more than 70%[20]. And to avoid the losses in the grid we could put the fuel cell in the van, where it would replace the normal rechargeable battery.

The first fuel cell was made more than 150 years ago and modern fuel cells are much more than delicate laboratory curiosities. They have been put into real experimental generating stations, real experimental vehicles on the road and communication satellites. Perhaps one day they will revolutionise both the electricity supply industry and the transport sector of the economy, but technologists do not use them in contemporary non-experimental stations or vehicles. Would we be more prosperous if they did?

Let us concentrate on the generating stations. Evidently if we are going to design them for maximum efficiency, we should not only stop building simple gas-turbine stations; we should stop building all ordinary (in Britain) coal-, oil- and gas-fuelled stations and go in for combined heat and power systems and fuel cells. But petroleum, coal and natural gas are not the only sources of primary energy; we also have nuclear energy and all the others to consider. The third misconception is that we can measure – or at least estimate – and compare the quantities of primary energy in different sources.

When technologists talk about the energy 'in' coal, oil or gas, what they have in mind is not a true physical property of the fuel but a partly conventional property, known as the 'calorific value'. It is defined as the quantity of heat (in megajoules) produced by burning one kilogramme of the fuel in oxygen; and it is a conventional quantity in that it depends on whether the products of combustion are cooled down to the ambient temperature or released into the atmosphere at above 100°C. The former condition yields a slightly higher value than the latter and they are known as the gross calorific value (GCV) and the net calorific value (NCV) respectively. For a typical black coal the difference is about 4%, for a typical fuel oil about 6% and for natural gas 10%.

The efficiency ascribed to a coal-, oil- or gas-fuelled generating station depends on whether it has been calculated from the GCV or the NCV of the fuel. For historical reasons, the published efficiencies of generating stations in Britain used to be based on the GCVs of their fuels whereas the corresponding statistics in the USA and many other countries were based on the NCVs. It does not require much imagination to envisage the kind of confusion which that difference sometimes created, particularly when neither the basis (GCV or NCV) of the statistics nor the data for the necessary corrections were published and the users of the statistics were unaware of the difference. Four or 6% can be very important in some contexts.

There were fierce debates about which value is 'correct', on the lines that it is not practical to cool the products of combustion down to ambient temperature so the NCV is 'correct'; however the calorific value is supposed to be a property of the fuel not the station, so the GCV is 'more correct', but such debates have no more depth than debates about the rule of the road – whether we should drive on the left or the right. Practical technologists were not much concerned because they use the calorific value mainly for specifying and testing the quality of the fuel which they buy and sell and burn. For those mundane purposes either convention will do, provided all the parties use the same convention, but of course the difference was a nuisance and the then Central Electricity Generating Board changed over to NCVs in 1982.

In the present context the essential point is that energy content is not a true physical property, even of coal, oil and gas. What about the other sources of primary energy? If you try to look up the efficiency of a hydroelectric station you will draw a blank because the energy content of a mountain lake or torrent is an elusive and not very useful concept to a technologist. Efficiency figures are published for nuclear stations but they exclude the efficiencies of the reactors[21], which convert the nuclear energy into heat, whereas the efficiencies of coal- and oil-fired stations, of course, include the efficiencies of the furnaces, which perform the equivalent function. So comparisons of the efficiencies of nuclear and non-nuclear stations are like handicap golf matches; they may be fun if you like golf but they obscure the relative proficiencies of the contestants. How much energy you can extract from those sources is so tied up with the technology adopted to extract it that there is no real point in trying to separate them. Coal, petroleum and gas apart, direct comparisons of primary energy are no more true comparisons of like with like than direct comparisons of the energy in petrol and electricity.

That is not to say that the usefulness of various sources of primary energy cannot be compared. One interesting, though by no means unique, way of comparing a uranium mine, a coal mine, a petroleum well (with or without natural gas), a tidal basin, a windy ridge, an area of sun-baked desert and some hot subterranean rocks would be to estimate how much electricity could be generated from each source. A cautious comparison of those estimates would be a legitimate comparison of like with like and it might serve a useful purpose, provided anyone who made the comparison kept firmly in mind that 'how much electricity could be generated' means no more and no less than how much electricity the designers of the stations would choose to generate from those sources. In other words we must understand how the designers think before we can make such a comparison. We cannot derive a set of rules for how designers ought to take decisions from the results of a study of how they have taken them or how the person who prepares the study for us believes they would take them.

Energy is one of the fundamental concepts of physics but essentially it is an abstraction. Abstract concepts are meat and drink to physicists but perhaps the rest of us should reflect on Alexander Pope's 'Essay on Criticism':

A little learning is a dang'rous thing;
Drink deep, or take not the Pierian spring:
There shallow draughts intoxicate the brain,
And drinking largely sobers us again.

Shallow draughts of the liquors of the energy spring are too intoxicating for judging whether an application of technology will add to our prosperity, and deeper drinking fails to quench our thirst for relevance. We need something less elusive.

THE COST METHOD

Coming back to the bled-steam boiler feedwater heaters in our coal-fired generating station, some little known facts about coal in Britain, a long time ago now, may clarify the issue and point us in the right direction.

Battersea power station, which was constructed in 1936, became a byword in Britain, but during and after the Second World War the quality of coal from British coal fields declined, for valid reasons, and Battersea had to burn, with great difficulty, a lower quality of coal than that for which it had been designed. Modern stations burn even lower-quality coal, which was an unburnable waste product until the process of pulverising it into tiny particles and blowing them into the furnace was developed during the 1920s and 1930s. So the early pulverised-coal stations were burning an otherwise useless by-product of mining high-quality coal. It would have been nonsense to instal more feedwater heaters or larger condensers or cooling towers just to save a waste product which would then lie about in ugly heaps in the countryside. That phase was short-lived because the new boilers soon used up all the surplus low-quality coal and it became the principal product of the

British coal industry. But the point is that the real gain, if any, underlying an improvement in efficiency, is the saving of a resource which is in some sense valuable. Sources of energy are normally valuable, which is why it is only common sense to keep efficiency in the forefront of the design process, but the paramount quantity in the design calculations must be the value, not the energy content of the fuel. Energy content can often be measured or estimated more easily than value and is often used as a surrogate for that reason, but it is a mistake to confuse them.

So what are the really relevant factors? As we increase the number of stages, we must acquire, accommodate and maintain more feedwater heaters and we must enlarge the high-pressure end of the turbine; these are the debit items. The credit items are the smaller low-pressure end of the turbine, the smaller condenser and the lower fuel consumption. We need a pair of scales into which we can throw the two debit items on one side and the three credit items on the other. None of the physical measures of the items – such as their number, weight, volume or energy content (however measured) – is appropriate, nor is any combination of them, and there are no overriding moral or human connotations because we do not intend to favour factory workers rather than miners or vice versa. There may be some environmental considerations but their relevance has been confirmed only recently and they are still very difficult to assess, so we shall leave them on one side until a later chapter. We are left with what we may conveniently describe as the 'cost method'.

The cost method proceeds by estimating the cost of every item affected by the decision. It must take account of when the cost will be incurred as well as the amount and it must embrace the effects of the decision on the other parts of the station. It works as well for choosing the materials for the conductors in the generator and the distribution system as for the number of feedwater heaters and all the other millions of decisions in the design process, not forgetting the initial choice of the type of station – nuclear, coal-fuelled, oil-fuelled or gas-fuelled, with steam turbines or gas turbines or both and so on.

Within an electricity supply system almost everything depends on almost everything else and the internal repercussions of many design decisions are complicated, but they can be safely left to the designers. Our concern will be with the external repercussions.

At some point in the argument, some enthusiasts for efficiency and energy conservation in general and electrical propulsion for transport in particular may try to reintroduce their ideas by the back door. They will concede that cost is important but then suggest that other things, such as energy conservation, are also important, sometimes more important, perhaps adding that people are not just economic animals, implying that cost is too sordid a criterion for important decisions. If prosperity itself is sordid – and some people maintain that it is, even when their own comfort and convenience are in the balance – what then should be the purpose of technology? Prosperity is the end for which the cost method is the means. What is the end for which energy conservation is the means? If the putative answer is protection of the environment, it remains to enquire whether energy conservation and environmental protection invariably go hand in hand, what we are to do if and when they do not and whether environmental protection necessarily conflicts with prosperity. In Chapter Nine it is argued that it does not.

Politicians, when they want to have their feet in two camps, and their administrators (on their behalf), often argue that they have multiple objectives in formulating their policies and that the community or the nation, as the case may be, should strive for energy conservation and minimum cost at the same time. That is like saying we will take the road to Glasgow and Edinburgh at the same time. The same road may lead to both destinations for much of the journey if it starts from a far distant origin, but eventually we must choose between them. Certainly technologists must and do strive for efficiency whenever and wherever energy is valuable, but not without regard to their consumption or utilisation of other valuable resources. Energy conservation *per se* will not bring prosperity. Conservation of valuable energy is no more than an aspect of the cost method – an aspect which is normally but not invariably important and never of over-riding importance.

Technologists seldom worry about why they use the cost method; they just use it, explicitly or implicitly. More about the details in Chapter Three. If they had to justify the habit, an important part of their argument might well come – perhaps paradoxically – from the sheer volume of the physical facts which they must bring into the calculations. Only the station design team can possess all the physical facts on designing the generating station and, by the same token, only the mine design teams can have all the physical facts they need to design the mines, and so on. The cost method is the only discipline which enables the teams to find their separate ways through their separate designs without endless, futile meetings to work out the effects of their decisions on one another. In short, the technologists' claim is that the cost method is the only rational, practical method of designing an application of modern technology, if its purpose is to contribute to the prosperity of the community which will use it. That is the special significance of money for technologists, which was mentioned in Chapter One (page 8).

Consequently the challenge to opponents of the cost method is that good intentions and high principles, however appealing they may sound, are not enough. Either the opponents must put forward a method which can reproduce the power of the cost method to weigh the factors involved in every technical decision numerically, and then submit to a comparison of the efficacy of their method with that of the cost method; or they must defend the sacrifice of prosperity which relinquishing the cost method entails. Some things are undoubtedly worth such a sacrifice, but many others are not.

REFERENCES AND FOOTNOTES

1) In Britain the approximate proportions in 1990 were: coal 68%, nuclear fuel 22%, oil and 'renewables' 10%. By 1996, the so-called 'dash for gas' had altered the proportions to: coal 44%,

nuclear fuel 30%, natural gas 19%, oil and 'renewables' 7%. The 'renewables' comprise mountain lakes, the sun and the wind, which are still only of minor importance in Britain.

2) In some small oil-fuelled stations the turbines are gas turbines, which are simpler than steam turbines, but the other thermal stations all have steam turbines. The latest large natural gas-fuelled stations have an even more elaborate combined cycle which uses both gas turbines and steam turbines.
3) The old, established meaning of 'generator' is 'machine which generates electricity'. During the privatisation of the electricity supply industry in Britain, a new and perversely different meaning came into vogue, namely 'company which undertakes to generate electricity for sale'. In this book the old meaning is retained. Those companies are referred to in Chapter Seven as 'generating companies'.
4) At the time of writing, research on super-conductors having no resistance is often in the news and they will probably have a profound effect on the design of generating stations and perhaps on the design of distribution networks in due course. But they will not affect the technologists' ways of thinking, which are what we are looking for here.
5) (For technologists). Ignoring the corrections for temperature and the skin effect, which are not important here.
6) A micro-ohm is one-millionth of an ohm.
7) Or tons. The Imperial ton is virtually the same as the metric tonne.
8) For example the Rosebery mine in Australia, which had published head grades of Cu 0.67% and Ag 137 grammes/tonne in 1983.
9) Electro-magnets are magnets whose magnetism is created by wrapping them with coils of conductors and passing currents through the conductors. In most large applications of electrical energy, including generators, the magnets are electro-magnets.
10) Nearly 900 times scarcer than copper, according to a consensus of estimates in 1983.
11) This anecdote may be apocryphal, but an article in *New Scientist* (January 1989) reported that Napoleon III sponsored the large-scale production of aluminium, that at the Paris Universal Exhibition in 1855 a display of the metal attracted as much attention as the crown jewels and that it was worth almost its weight in gold.

12) For example Drax generating station, designed by the then Central Electricity Generating Board and completed in 1986.
13) Physicists and technologists will recognise this corollary of the Second Law of Thermodynamics.
14) Another corollary of the Second Law.
15) John Billings' *Encyclopaedia of Wit and Wisdom*, 1874.
16) Just as mass is measured in kilogrammes. The physical definition of a joule is complicated unless one is already familiar with it, and in the present context the precise definition is not important. The familiar kilowatt-hour of electrical energy is 3,600,000 joules.
17) 'Petroleum' is the correct term for the liquid which comes out of an oil well. It is often referred to as 'crude oil' or 'oil', but 'oil' may also mean one or other of the numerous products of a petroleum refinery. It is therefore clearer to write 'petroleum', although perhaps it sounds pedantic.
18) 'Energy consumption of electric, petrol and diesel light goods vehicles in London', *Transport and Road Research Laboratory*, LR 1021, 1979
19) The large modern natural gas-fuelled, combined cycle stations mentioned in the above second footnote have efficiencies of up to 51%, but that technology was not commercially available when this comparison was made.
20) *Handbook of Fuel Cell Technology* by Carl Berger, 1969.
21) Nuclear energy is produced by converting mass into energy at the rate of 100 billion megajoules a kilogramme, but in a reactor only a minute fraction of the mass of the nuclear fuel can be converted, so the efficiencies of the reactors would be minuscule on that basis, and there is no other relevant basis.

3

THE TECHNOLOGISTS' SECOND METROLOGY

> Mr Lely, I desire you would use all your skill to paint my picture truly like me, and not flatter me at all; but remark all those roughnesses, pimples, warts and everything as you see me, otherwise I will never pay a farthing for it.[1]

So spoke Oliver Cromwell to his portraitist. This chapter, which portrays 'Oliver Cost-Method', has been painted in the same spirit of 'truly like me', including the warts.

Technologists use the cost method right across the whole range of applications of technology, from nuclear generating stations, aeroplanes large and small, factories, chemical process plants, hospitals, ships, ports and airports, railways and roads, rail and road vehicles of all kinds, medical and surgical equipment (which is nowadays highly technological), computers of course, down to domestic boilers, dishwashers and hair dryers. They are all called 'projects' while they are being designed and manufactured or constructed – not a very elegant term perhaps, but sufficiently descriptive.

The essence of the cost method is that the designers, having decided what the product or service is intended to provide, try to minimise the cost – in a defined sense – of doing so. The definition of the cost to be minimised is an important part of the metrology, just as the definitions of length, mass, time and other physical quantities are important parts of physical Metrology. In both cases the definitions are not as obvious as they may first appear and getting them right has profound effects on technology, physics and other branches of science.

Very often there is a balance to be struck between cost and quality and in the early days of any new technology, whether it be electricity supply, air transport, information technology or any other, the cost measurements may be very inaccurate and the balance correspondingly precarious, just as the physical measurements in the early days of the science of nuclear fission and nuclear fusion were very inaccurate. But in both science and technology the metrology eventually improves as and when the scientists and technologists perceive the necessity or desirability of improving it.

The cost method is not synonymous with the so-called 'price mechanism', which supposedly regulates the prices and quantities of goods and services automatically in a free market economy. The cost method is very much a conscious procedure, not an automatic mechanism.

To assent to the technologists' tacit assumption that the cost method is the only rational, practical method of designing their projects is much more radical than to grant that it is a very good method, or even that it is the best method in particular circumstances. For in allowing technologists to use that method we are consenting to their wielding a very powerful tool, and it is not a tool which automatically produces the desired results by itself. It has gradually changed, and will continue to change, the economic structure of society, building up industries here, reducing and sometimes destroying them there, but whether the results are beneficial or harmful depends on the institutions and features of the society in which the tool is used. In formal language, the use of the cost method is a necessary but not a sufficient condition for deriving prosperity from technology.

An important further condition was partly visible below the surface of our first simple example in Chapter Two. Using the cost method, the generator designer may not choose copper for the conductors if its price per kilogramme is the same as the price of silver. In that case, the cost of 50 silver conductors, weighing 5 tonnes, would be only 16% greater than the cost of 50 copper conductors, weighing 4.3 tonnes, so the designer would probably choose silver for the benefit of the fuel saving. Common sense would dictate that the prices of the two metals must differ enough to ensure that generator designers always choose copper, not silver – except perhaps in extraordinary wartime circumstances when the objective is no longer prosperity but something like victory or short-term survival, in which case the cost method may be inappropriate or require radical modifications. But the condition is much more complicated than that. The price of copper affects not only whether it is used instead of silver but whether it is used instead of aluminium and, even more to the point, how much copper is used – the length, cross-sectional area and number of conductors. In those decisions, the prices of aluminium and of all the other components of the generator, which depend on the sizes and number of conductors, will also enter the calculations. Moreover the price of copper affects the design of the copper mine and how much exploration the mining company undertakes.

In general, the necessary concomitant of the cost method is a rational, practical system of pricing and methods of charging, with the specific purpose of linking the technologists' decisions to the locations at which the benefits of those decisions are expected to accrue. It is a dual purpose in that the price of the commodities or other resources must guide both sides of the transactions. Our interest in the method therefore has two strands: to understand how technologists obtain and compare the costs on which they base their designs, and to derive some of the features which the prices and methods of charging for the products and services of technological applications must have in order that the cost method, in the hands of the designers, will bring prosperity to the community.

This chapter is concerned mainly with the former strand. Our examination of prices and methods of charging has to be postponed until Chapters Six and Seven, but meanwhile it is worth emphasising that the costs of a project are not just a nuisance which must be taken into account in its design. They are the signposts along the only path by which the project can bring prosperity to the community which engages in it. The path is rough in places, but it cannot be circumvented.

SIX ASPECTS

The cost method has six main aspects. It encompasses the means of *obtaining estimates,* not only of the costs of projects but also of the options which the designers will eventually reject. It allows for *payments at different times* during the construction and operation of projects and for *monetary inflation* during those periods. It can distinguish, where necessary, between the *two kinds of cost* mentioned in Chapter Two. It can deal with most of the *uncertainties* inherent in plans for the future (though not, of course, eliminate them), and finally it encompasses the *execution and completion* of contracts. In the present state of the method, it does none of these things perfectly, but it does not – or at least it need not – omit any. Astute critics will have noticed that taxation is not a separate heading. That is because it is more of an additional complicating factor than a separate aspect. It crops up in connection with two of the foregoing aspects in this chapter and again in Chapters Four and Five.

Some of the aspects of the cost method can be illustrated adequately by examples in electricity supply, but not all of them, and in any case some variety in the illustrations is desirable. So from time to time we shall turn to other types of project and in particular to a study some years ago of the future prospects for electric road vehicles, so called, in Western Europe[2]. A few words about that study are in order before we come to the six aspects.

'Electric road vehicles', or 'EVs' for short, became, for a time at least, the internationally accepted term for one restricted class of vehicles, namely electrically propelled road vehicles with batteries and no other power supply when the vehicles are moving. In ordinary parlance, electric trams and trolley-buses are also electric road vehicles, but not in the acquired parlance of the participants in the study and the manufacturers of the vehicles, nor the transport departments of universities and governments in Western Europe and many other countries. The full specification, 'electrically-propelled . . . moving', would have been much too much of a mouthful; nobody really knew what to call them, so the fairly short label stuck.

EVs in that restricted sense have a slightly longer history than road vehicles with petrol (gasoline) engines and much longer than vehicles with diesel engines, both of which in this context are called 'conventional' vehicles. But for many years very few EVs had been seen on the roads, except for the slow milk delivery vans in Britain, whose numbers were declining at the time of the study. The four main reasons for this sparseness were:

1) a reliable battery was about 150 times as heavy as a tank of petrol or diesel fuel which would give a conventional vehicle the same range before the tank had to be refilled, or the said battery exchanged or recharged;
2) recharging the batteries took several hours rather than the few minutes needed to refill a fuel tank;
3) the batteries were expensive and wore out before the vehicles; and
4) the vehicles themselves, without batteries, cost three to five times as much as their conventional counterparts, i.e. with approximately equivalent capacity and performance.

On the other hand:

1) EVs were much quieter;
2) they did not pollute the air in towns;
3) lighter batteries with potential useful lives of up to four or five years had been developed and more improvements were expected;
4) both the vehicles and the batteries would be much cheaper if they were produced in large numbers; and

5) the EVs themselves did not require for propulsion any product of petroleum, of which Western Europe was a net importer, because electricity can be generated from several different sources of primary energy, as mentioned in Chapter Two.

The main questions for the study were: could EVs contribute to the prosperity of Western Europe and, if so, what, if any, actions by governments – in the way of legislation or policy co-ordination – would facilitate that contribution? We shall come to the study's conclusions and recommendations in Chapter Nine; in this present chapter our concern is solely with its use of the cost method.

If EVs were to contribute to Western European prosperity, they would have to escape from their 'poverty trap', in which they were produced in small numbers so they were expensive, so very few were sold so they were produced in small numbers. The way out was for the participants was to ask themselves and their backers a series of questions as follows:

1) 'Roughly how many EVs, and of what types, might be physically capable of replacing conventional vehicles if costs were temporarily ignored?' After a period of estimation and calculation, the answer was: 'Six million small cars and one million small delivery vans, all on short journeys, if...'. But the provisos, though important in the study, need not delay us here.
2) 'What would be the service lives of these vehicles?' Answer: 'Six to twelve years, depending on their annual mileages.'
3) 'So with, say, four assembly lines for each class of vehicle, that would mean line production rates exceeding 100,000 small cars and 10,000 small vans a year?' – 'Yes.'
4) 'Then what would those EVs and batteries cost to make and sell in such numbers?'

The study nearly foundered on that last question, for two reasons to which we shall return, but estimates eventually came through from some well-disposed manufacturers. The participants in the study had then to estimate the overall costs of acquiring and operating those EVs, at various annual mileages, compared with the corresponding costs of their conventional counterparts. Some costs, such as that of the driver, could be omitted because they would be the same for either type of vehicle with any given pattern of use. The costs which were different were of:

1) the vehicles – with one exception, the EVs were still dearer but had longer lives;
2) the propulsion batteries – the conventional vehicles of course had none;
3) energy – electricity or petrol or diesel fuel;
4) other (minor) operating costs – which favoured the EVs.

We shall return to those estimates in a later section of this chapter.

OBTAINING THE ESTIMATES

In Chapter Two, we broke into the middle of the design of a generating station and looked at two decisions – the material for the generator conductors and the number of stages of bled-steam boiler feedwater heating. Although they are both a long way from the start of the design process, each of these decisions affects other parts of the design. Thus the number of feedwater heaters affects the sizes of the high-pressure and low-pressure ends of the turbine and of the condenser and cooling towers, also the boilers and to a slight extent the coal-handling and ash-handling plants; so we should ideally like to know the costs of all these items for different numbers of heaters as

well as the costs of the heaters themselves before we decide that number. Similarly, if the relative prices of copper and aluminium were to alter enough to make aluminium conductors a real possibility, the choice between copper and aluminium would affect the number and dimensions of the conductors and their insulation, the amount of iron to provide paths for the magnetic fields, the size and weight of the generator, the strength of its foundations and the size of the building, also the quantity of cooling water for the conductors and the sizes of the pumps, motors, cables and controllers to circulate it; so we have another list of costs we should like to know. Both lists peter out in the end but they are long enough to pose problems about where the information is to come from.

Unfortunately those lists tend to become longer, and the information more difficult to obtain, the nearer we come to the beginning of the design. Nowadays stations are seldom conceived in isolation; normally there is a grid supplying a large area (often a whole country) and a few dozen stations in operation[3], some nearly new, most not so new. Here is a list of some early decisions:

1) When shall we need some new generating capacity? Our total demand is growing slowly, albeit irregularly, and our old stations are gradually deteriorating and becoming uncompetitive with newer stations (more about those two processes in Chapter Four), so we shall need some new capacity sooner or later, but the timing is important because generating stations are expensive.
2) Should we build large stations with large generators infrequently or smaller stations with smaller generators more frequently? Large stations are cheaper per kilowatt of capacity, but their output has to be distributed more widely, which involves more energy loss and expense.
3) What kind of station(s)? Our total demand fluctuates between night and day and between summer and winter, so do we need a new 'base-load' station, i.e. one which will operate continuously (perhaps a nuclear station), or one or more 'peak-load' stations, which will operate for

only a few hundred hours in any one year on cold winter mornings (in Britain), or something in between, which will operate on base load when it is new and then be gradually relegated to peak-load operation? A base-load station can have a much lower fuel cost per kilowatt-hour of energy output, but the station itself may cost up to three times as much as a group of low-efficiency peak-load stations of the same total capacity, which need not all be in one place.

4) Where shall we build the station(s)? On or near the coast where sea water or estuary water may be available for the condensers (thus saving the cost of cooling towers), or near a coal field, an oil well or a natural gas well, or near the largest aggregation of demand, where land is probably expensive and possibly less suitable for supporting the heavy plant and equipment, but the additional load on the grid would be smaller?

To answer the very first question we should like to have estimates of how much the stations and their fuel, stores and staff will cost, and for those estimates we should like to know whether the stations will be large or small, what kind of stations they will be and where they will be built. In other words we should like to have estimates of construction costs and operating costs over the whole range of possible types, designs and locations of stations before we can take our first decision. But clearly we are not going to get information in any such degree of detail at that stage and we shall have to make a start without most of it; moreover the best information we can lay our hands on is likely to be vague. In general, an important part of a designer's skill is in making the best use of sparse uncertain estimates, often amounting to little more than educated guesses. Those inescapable uncertainties severely limit the contribution to prosperity which even the most brilliant designer can achieve.

The uncertainty of an estimate depends of course on the quality of the data from which it has been calculated. At any stage in the design, some of the data will come from inside the organisation which is involved at that stage and some from outside. For example a turbine manufacturing company can calculate the effects of making larger or smaller turbines on the utilisation of its existing machincry, factory space and labour force (including its designers) and must devise a system of allocating the costs of those items between the turbines it will produce. If that manufacturer also makes feedwater heaters, the same will apply to the heaters and the cost allocation system will be correspondingly more complex. It will have to cater for producing turbines of the same output with different numbers of heaters to suit different customers with different fuel costs, and for producing turbines without any heaters and perhaps heaters without any turbines on some occasions. But there is always a boundary, outside which the organisation must depend on commercial transactions for cost data, and in the long run all the data must come from external transactions.

It is not that internal data are inherently better or worse than external data. Internal data include the known costs of machinery and factory space already acquired and the known rates of pay for an existing labour force, but the allocation system is inevitably uncertain because it depends on the future numbers and sizes of the turbines and heaters to be produced. Allocating external costs is simpler because they are incurred in respect of specific items, but then the costs themselves are often more uncertain until they are actually incurred, by which time the manufacturer will normally be committed to a price.

There are broadly four sources of external supply and therefore of external data: namely:

1) open commodity **markets**, in which the prices are the outcome of competition between sellers and between buyers;
2) **monopolies**, which decide their own prices but are often regulated in some degree by the licences or obligations under which they are constrained to operate;
3) **oligopolies**, where there are not enough suppliers to bring about genuine competition; and

4) contractors for special goods and services for each project, whose prices are the outcome of **competitive tendering**. We will take them in that order.

For a generator, the metal for the conductors will probably be acquired in a metal market or through a manufacturer whose prices are governed by the prices in that market. Designers can study or take advice on the trends of those prices and so estimate the price at which the conductors can be purchased when the generator is manufactured. In the long run (ten years or more) the prices will affect the number and sizes of the conductors, but changing them has expensive repercussions on the sizes and settings of the machines for cutting and shaping the conductors and the other affected parts, so such changes are usually postponed until a major development is under way, such as a larger generator or a new method of cooling or both. Changing the material, e.g. from copper to aluminium, would of course constitute a major development in its own right. Nevertheless the market mechanism is effective in regulating design changes and directing them towards more prosperity, or minimising the loss of prosperity caused by external influences, provided that mechanism is not interfered with and the prices are published. There is then no great difficulty in obtaining the data for the design calculations.

In general price fluctuations (as distinct from trends) are ignored in the calculations because, even if they could be accurately forecast, it would be too expensive to make repeated changes to accommodate them. They sometimes affect the timing of projects in which the cost of the commodity in question is a large part of the total cost (conductors for generators are not in that category), but with two exceptions fluctuations in commodity prices are irrelevant to the contribution of technology to the prosperity of a community. The first exception is when the commodities are bought and sold across the boundaries of communities, in which case of course a price increase favours the selling community at the expense of the buying community, and vice versa for a price reduction. The second is when the fluctuations are mistaken for long-term trends and so cause economic projects to be redesigned or abandoned or uneconomic projects to be undertaken. Price fluctuations are bound to affect the prosperity of individual companies and undertakings but those effects are mainly transfers of wealth from one organisation to another; the gains balance the losses. It is only when a price change affects the designers' decisions for one or both parties to a transaction that it affects the prosperity of the community in which it is made. For that reason the distinction between a fluctuation and a trend depends on the complexity of the manufacturing process. A large increase in the price of copper might perhaps lead to aluminium being used instead of copper for domestic water pipes, even if it was expected to last for only a few years, but it would not affect the design of generators until the designers were convinced that it was going to persist for several decades.

For commodities, such as fuel and stores, which are required throughout the service life of a project, the time scales are correspondingly longer. The designer of a generating station is concerned with the prices of alternative fuels up to 25 years ahead, for which even the best estimates are bound to be speculative, but fortunately the method of taking account of payments at different times reduces the element of speculation in the design – as will become apparent in the next section.

The main problem with monopolies is how to regulate their prices and practices without weakening their motivation to perform efficiently, but their regulated prices can be stabilised over comparatively long periods, thereby giving the users of those goods and services more confidence in their future costs than they can get in an open market. The corresponding problems with oligopolies involve balancing the advantages and disadvantages of limited competition versus regulation. Both sets of problems are tackled in Chapter Seven.

The advantage of confidence that a set of prices will not rise unexpectedly is fairly obvious. One knows that projects which will use those goods or services will not become uneconomic after the investments in them have been made. It might be thought that any reduction in the price of a good or service, expected or not, would always be welcome, but it is not if plant and equipment have been installed to reduce the consumption of that good or service, for example if a generating station has been constructed (perhaps utilising surplus steam from a chemical process) to reduce the consumption of imported electricity and then a reduction in the price of electricity makes the project uneconomic. However the value of stable long-term prices has to be judged against the cost to the supplier of absorbing fluctuations in its own external costs. Sometimes it is cheaper in the long run to design a process (at some additional capital cost) to operate whenever the good or service it requires is cheapest from time to time than to pay a premium for a stable long-term price. We shall touch on those questions in Chapter Six.

Coming now to the prices of contracts obtained by competitive tendering, anyone who has tried to add something like a garage to his or her house will know that it is not easy to obtain realistic quotations from several contractors. That is because preparing a quotation, even for a simple thing like a garage, is an expensive business. It involves visiting the site, measuring, working out the costs of materials, labour and overhead costs, judging how much profit to add and typing the offer. At the other end of the scale a slight impression of the amount of work involved in tendering for a nuclear generating station may be conveyed by recounting that each tenderer requires a fleet of vans just to transport the documents to the purchaser's premises. And those documents are no more than the visible snout of the hippopotamus in the lake; below the surface the bulk of the animal comprises teams of designers, estimators, computer programmers and operators, clerks, word-processor operators, etc., not only at the tenderer's premises but also at the premises of a number of intended sub-contractors, who will supply items and do work which the main contractor is unable or unwilling to undertake. Tenders for such things as EVs, hospitals and ships are at intermediate positions on the scale.

All along the scale, the unsuccessful tenderers must recover the costs of preparing their tenders from their profits on other contracts for which they have tendered or will subsequently tender successfully; otherwise they will go out of business. So naturally tenderers are reluctant to prepare tenders unless they can scc a good chance of a profitable contract[4]. If each tender had also to include different prices for all the options the purchaser wished to consider, the whole competitive tendering process would become impossibly expensive and would break down. Enquiries for competitive tenders have therefore to be restricted, one way or another, to projects or parts of projects, which have already been designed in enough detail to provide a single specification for each enquiry. It then follows that the only ultimate source of valid data for estimates are the prices of accepted tenders for previous projects – in effect the sums which change hands in actual transactions, of which there are very few compared with the number of estimates used in the course of preparing the single specifications. The prices in the other offers do not represent valid data (except to the extent that, if they are close to the accepted prices, they confirm that those prices are reliable) for the very reason that they are for the same specifications and have been rejected.

This brings us to the essential differences between an estimator's view of the commercial transactions in an application of technology and the view of an accountant preparing a set of accounts for an annual report. The functions of annual accounting in that sense are: compliance with the relevant legislation, prevention of fraud and embezzlement and calculation of the annual turnover, the receipts and expenditure under prescribed headings, unrelated to particular projects, the profit or loss and the values of the assets and liabilities, again under prescribed non-project headings, at the end of every accounting year. Those functions are quite different from the functions of ascertaining the costs of individual projects and parts of projects and estimating the costs of options for future projects, and they require different tenets and rules. In short, annual accounting is concerned with data aggregation; project estimation is concerned with data analysis. That is not to imply that accountants cannot or do not engage in the latter kind of activity, but that the approach they habitually adopt in annual accounting is not well suited to the functions of project estimation. In this chapter the two functions are separate, but in the next and subsequent chapters they will be seen to overlap.

Sometimes an enquiry will offer a bonus for efficiency in a test to be conducted on completion instead of specifying every feature of the design in detail, thus leaving the tenderer to calculate whether to incur extra cost, for example by offering a larger condenser, so as to receive a bigger bonus. But then there must be a penalty for not achieving the offered efficiency and in practice the penalty can seldom be large enough to compensate the purchaser adequately for the non-achievement. Consequently, although efficiency bonuses and penalties are normal features of contracts in which efficiency is important, they are not normally used to decide the major features of designs. Sometimes a specification will call for optional prices for a few minor features, or they may be offered voluntarily by the successful tenderer, but the emphasis has to be on the word 'minor'. For the major decisions, the costs of the options must come, directly or indirectly, from previous projects.

In the prosperous industrialised countries, competitive tendering to purchasers' specifications is the established method of regulating the prices of transactions whose specialised nature precludes their being conducted in an open market. Few technologists in those countries would want to abolish competitive tendering, although it puts a strain on the financial resources of the tenderers, but it also tends to restrict the availability of reliable data. That disadvantage could be lessened if the practice of publishing tender prices became the norm. The purchasing departments of public authorities often adopt that practice to eliminate or reduce corruption, but it also provides data for other suppliers and other purchasers of similar plant and equipment. It is commonly opposed in the private sector because information is always valuable[5], but it should be judged on whether it increases the contribution of technology to prosperity, and on that count any increase in the availability of valid data is beneficial.

PAYMENTS AT DIFFERENT TIMES

A generating station, like any largc projcct, is paid for during the period of its construction, generally two or three years but sometimes longer. Its operating costs – mainly fuel, staff and stores – are paid at intervals ranging from daily to monthly, sometimes quarterly, throughout its service life, which is typically 25 years. As the station becomes older, its efficiency declines slightly and it is overtaken by newer stations with higher efficiencies emanating from continuing technological progress and/or the availability of cheaper fuel, so its annual output is commonly reduced to give preference to the newer stations, which of course reduces its operating cost, year by year. Similarly transport vehicles are generally paid for when they are acquired and their operating costs are also paid at intervals ranging from daily to monthly during their service lives, which range from five to twelve years or so, but EVs also require new batteries every two to five years, depending on their annual mileage.

The questions in this section are: how can the designer of a project such as a generating station combine those various payments to compare the estimated costs of a station with those of a similar station with some design changes, such as some more bled-steam boiler feedwater heaters or a larger condenser or cooling tower, or all three. And correspondingly in the transport example: how can the costs of an EV be compared with those of its conventional counterpart on the same missions?

Sometimes the user of a transport fleet rents or leases the vehicles and/or their batteries from a financing organisation, which actually pays for those items, but such an arrangement merely moves the question to the renting or leasing organisation, which must also use the cost method if the best decisions are to be taken. For simplicity we shall concentrate on the case where the owners are also the operators and pay directly for the goods and services they require, as and when they acquire them.

Superficially the simplest way of combining the payments at different times would be to add them all up over the service life of the project and call the total the 'whole-life cost' of the project, but there is an objection, which any economist will confirm. It is that, other things being equal, we prefer to pay as we go rather than in advance. Consequently we find that we can lend any money we can save at interest or invest it for a profit until we spend it, or we have to borrow money at interest or issue shares to pay for anything in advance, which reinforces our preference. Moreover there is always a risk, which may be large or small but never quite zero, that something will go wrong and we shall lose some of or all the money we spend in advance, or we shall not get as much benefit as we expected from that advance spending.

Technologists fully understand and share that preference. Its effect is that, if the whole-life cost, so calculated, is the same for any two options, they choose the option which minimises their advance spending. If the cost of the fuel saved by an additional feedwater heater, which will be felt gradually over the next 25 years, were no more than the net cost of having that additional heater, which must be paid now, the designers would choose without hesitation not to have it. The additional heater must save substantially more fuel than that before they will begin to consider it. Similarly, unless transport operators are enthusiastic about EVs for their own sake, or believe they have some virtue which cannot be expressed in money, they will not begin to think of buying them until they can estimate that their operating cost savings, to be achieved over ten years or so, will amount to substantially more than their additional acquisition costs. In both cases the question is: how much is substantially more?

The payback-period method

There are two main ways of bringing that preference into the calculations. The first, which is known as the 'payback-period method', is simple and widely used; indeed many managements insist on it, but it is crude.

Suppose we are designing a generating station with a total output capacity of two million kilowatts, for which some proposed design changes, comprising perhaps an additional feedwater heater and a slightly larger condenser and cooling tower for each of four turbines, would save fuel costing about £100,000 a year. If the proposal would cost £500,000, the fuel saving would, in a sense, 'pay back' that cost in five years, which is therefore called 'the payback period'. So instead of wondering how to balance £500,000 now against £100,000 a year for the next 25 years, we could try to decide whether it would be worth the risk of being out of pocket for five years, and either paying an unspecified rate of interest or foregoing some unspecified profit from other investments meanwhile, in order to save £100,000 a year in the sixth and subsequent years. The theory of the method is that, if we have a budget for such proposals, we can calculate the payback periods of all the ones we have in mind, arrange them in order of increasing payback periods and, starting from the top of the list, include as many as we can afford out of the budget; but the theory provides no guidance on the size of the budget. Or we can specify a so-called 'cut-off period'[6] and include only those proposals with payback periods less than the cut-off period; but again the theory does not specify the cut-off period. It is commonly stipulated to be between two and four years, which would rule out this particular proposal.

The payback period can be recalculated if the annual output will be reduced year by year, and allowance can be made for the risk of failing to achieve an intended improvement by stipulating a shorter cut-off period for new developments than for well-tried features, but those are the only virtues of the method, apart from its extremely simple arithmetic. Its crudeness lies in its incompleteness (in leaving the budget or the cut-off period to be stipulated arbitrarily) and its failure to bring the service life of the project and the rate of interest (or expected profit) at which it will be financed explicitly into the calculation. The omission of the service life is especially crude in long-life projects because the savings continue long after the payback period has passed. The method has probably led to more and larger missed opportunities for deriving prosperity from technology than any other single cause. We shall come to some examples after considering the alternatives.

The discounting method: the present value

If the payback-period method were the only way of combining payments at different times, it would be worse than an ugly wart on the face of Oliver Cost-Method. An uncommitted observer might wonder whether technologists ought to continue using the cost method after all. Fortunately there is another, much more rational, way of doing what we want. Variants of it have been used in the electricity supply industry through much of its history, and elsewhere, under different names, the shortest of which is 'the discounting method'.

The discounting method comes from the notion of compound interest, which is taught in schools but is then commonly discarded in later life or used only with reluctance, even by many technologists, although they regularly use much subtler concepts in their professional lives. Why people who have pocket calculators with all the relevant functions and who are happy to calculate such quantities as the kinetic energy of rotating masses, the bending moments of beams or the Fourier components of complex waveforms, and often to make far more complicated calculations than any of those, should jib at discounting calculations is something of a mystery, but they do. However it can be hoped that more and more of them will acquire the habit of pressing a few buttons on their calculators from time to time to get the advantages of the discounting method. A computer is not essential, except for the more elaborate applications, but computers are so cheap now that they are nearly always used.

This book is not a mathematical treatise, so we shall not try to derive the discounting formulae rigorously from first principles in the main text. Suffice to say that, if £100 is borrowed at, say, 7% compound interest, the amount owing will grow at an ever increasing pace. It will be £140 after five years, £197 after ten years and so on. So, in a project which is financed by borrowing money at 7% interest (or foregoing other profit at that rate), £100 now has the same value as £197 in ten years time. Putting it the other way round, £100 in ten years time has the same value as £(100 × 100/197) = £51 now. In the vernacular of the method the 'present value' of £100 ten years hence is £51 if the 'discount rate' is 7% a year. The discount rate encapsulates our preference for paying as we go. It is the rate at which future savings (or payments) are to be reduced in value for the purpose of comparison with payments (or savings) at an earlier time, usually now.[7]

From that beginning, we can calculate the present value of any stream of savings we care to envisage. We do it by first calculating a 'discount factor' for every year throughout the service life of a project. For example, if the discount rate is 7%, the discount factor for the tenth year is 0.51 (100/197 as above). For the twentieth year it is 0.26 and so on. Then we multiply the saving in every year by the discount factor for that year and add up all the products to get the present value of those savings.

That may seem like a very laborious operation and it was before electronic calculators were invented, but they have revolutionised that situation. If the stream of savings is at a constant rate, we can use a formula for the whole calculation – rather like the formula building societies use to calculate mortgage payments. If the savings diminish by the same percentage every year, the formula is a little longer. If the savings are irregular, the calculations are more laborious, even with a computer, but irregularities are difficult to predict and small irregularities do not make much difference to the answer, so that situation does not often arise. For readers who are curious to learn the details or wish to refresh their memories, the formulae and some examples are given in Appendix 1 of this chapter.

In the foregoing example, with a discount rate of 7%, the present value of a stream of savings at the constant rate of £100,000 a year for 25 years is £1.17 million, so it would be well worth while to invest £500,000 now to obtain that stream of savings. It represents a gain of £670,000 (£1.17 million – £500,000) – even after we have allowed for our reluctance to spend money in advance. Spending money in advance is normally an essential part of the process of prospering from technology.

If the discount rate is 12%, the present value of those savings falls to £780,000, which is still amply large enough to justify an investment of £500,000. However if the discount rate is 20%, the present value is only £495,000, which is less than the £500,000 we should have to invest to obtain those savings. We should do better not to make that investment.

Coming now to the risks, there are mathematical techniques for assessing the risks of occurrences if there are good records of the past frequencies of those classes of occurrence. The best known example is the risk of death in peace time, which insurance companies can assess with enough accuracy to enable them to calculate their premiums, instead of just guessing them. Sometimes very tiny risks can be calculated by combining other less tiny risks; for example the risk of a boiler bursting can be calculated by combining the risks of a rapid, uncontrolled rise of steam pressure and of the failure of any one of its several safety valves, for which separate frequency records can be accumulated in the course of normal operation and testing. (And then the risk can be reduced by increasing the number of safety valves.) But there are seldom any objective data on the financial risks of design changes (as distinct from the physical risks), so one is usually reduced to educated guessing, based on previous general experience with that type of design feature, the results of development tests and so on. The important thing is not to bury the guess in the calculation but to keep it in view until the end, so that its effect on the consequent decisions can be seen. Then if a decision seems to be too finely balanced, the effect of changing the educated guess can be brought out, which may push the decision firmly one way or the other.

Probably the commonest and certainly the simplest way of inserting an educated guess is to put into the calculation a discount rate which is larger than the rate which would apply if there were no risk, for example by using 12% instead of 7%. Then when nothing goes wrong the value of the operating-cost savings, discounted back to the beginning of the project (now in the past), will turn out to be greater than was expected (£1.17 million instead of £780,000). We can imagine the surplus going into a fund, which will compensate for the occasions when our expectations are disappointed. The contributions to that imaginary fund of the items with long service lives will be greater than the contributions of shorter-lived items, which satisfies our instinctive feeling that the risk increases with the length of time an investment is at risk.

That does not mean we should always use a high discount rate to avoid any risk of losing part of our investment. If we pitch the rate too high, we shall miss numerous opportunities for profitable investment. Our object is rather to balance the gains and losses to and from the fund in the long run. However it would have to be an imaginary fund. It would be idle to pretend that enough data could be obtained to set up and monitor a real fund for such a purpose. Financial risk evaluation is an art, not a science, unlike physical risk assessment, which is gradually becoming a science.

The specific risk of failing to achieve an intended operating-cost saving by increasing the investment in a project is only one of the uncertainties which designers must face. The other more general uncertainties are dealt with more conveniently after considering the difference between the two kinds of cost.

The discounting method: the equivalent annual cost

So how broad is the reach of the discounting method? Is it restricted to testing minor changes in a project whose main shape has already been decided? No, the present value can also be used to compare the costs of two completely different ways of producing or providing a good or service, with the proviso that they must start at the same time and have the same service life. Formally the objective is still to minimise the total present value of all the costs of the project, including the investment costs of the plant, equipment and buildings. But a variant is necessary if the service lives differ, because then the present values are not directly comparable. Historically the variant is older than calculating the present value, but it fell out of fashion and its virtue in this particular circumstance is sometimes forgotten.

For example, in the European EV study, the total present value of all the cost items of an EV with a service life of ten years cannot be simply compared with the corresponding total present value of its conventional counterpart with a service life of only eight years, because a new conventional vehicle would be required while the EV was still able to give two more years of service. Moreover the service life of an EV is not generally an exact multiple of the life of its battery. The way round the difficulty is to turn the formula upside down! Or perhaps, for historical accuracy, one should talk about turning it the right way up!

We can think of the present-value calculations as piling up all the operating costs on to a heap on top of the investment cost or costs at the start of the project, with the rule that the later a cost has to be met, the smaller is its contribution to the height of the heap. In the variant, the acquisition costs of the vehicles and the batteries of the EVs are spread evenly over their respective service lives, on top of the operating costs, like butter and jam on a slice of bread. The rule is then that the shorter the service life of an item the thicker the butter or jam must be. We can call those thicknesses the 'equivalent annual cost' of the vehicle and its battery (if it has one) respectively, and define that term formally as the 'uniform annual cost' over the service life of an item having the same present value as the acquisition cost of that item[8]. The overall thickness of the bread (representing the annual operating costs, which must be constant for this variant to work), plus the butter and the jam for the EVs, then provides an objective measure of the overall costs of the project – no less objective than the present value when all the items have the same service life. The formula is given in Appendix 1.

Some readers like words; others prefer a few figures for ease of comprehension. For the latter category, here are some results of one of the hundreds of comparisons which came out of the study. The two vehicles were an electric van with a payload of a little over half a tonne and an approximately equivalent petrol van, both running 35 kilometres a day throughout a working year of 250 days in Sweden in 1985. The acquisition costs were 10,000 European Currency Units, or ECUs[9], for the EV, 1,200 ECUs for its battery and 8,600 ECUs for the petrol van. The service lives (estimated of course) were 10 years, 4.4 years (for the battery) and 8 years respectively, and the discount rate was 5% a year. The annual costs, all in ECUs, were:

	Electric	Petrol	Advantage of EV
For energy + maintenance + tax	330	820	490
Equivalent annual cost of vehicle	1300	1330	30
Equivalent annual cost of battery	310	–	–310
Totals	1940	2150	210

That is to say that that particular EV under those particular conditions would be 210 ECUs a year cheaper to buy and run than a petrol van on the same missions.

A slight presentational disadvantage of calculating equivalent annual costs is that they may give the impression that they will appear like that in the annual accounts, especially when they are set out in a table. It is even asserted sometimes that they can be resolved into two components, the interest and the annual depreciation, but that is a mistaken view. In designing a project to produce or provide a specified good or service, it is not necessary to take any view of where its various costs will turn up in the annual accounts. Nobody supposes that when the present value is calculated, the values of the items of operating cost will appear in a heap or list in the first year of the annual accounts and equally there is no valid reason to suppose that the acquisition costs will appear as equivalent annual costs. Both calculations are devices to facilitate comparisons of the estimated future costs of projects, which are not the concern of annual accounts, as we shall see in the next chapter.

The equivalent-annual-cost variant covers the circumstance that items have different service lives, but it cannot deal with varying operating costs, and neither technique, as described, can deal with projects starting at different times. How then does the discounting method manage when the items have different service lives and varying operating costs in the same project and there are many co-ordinated projects starting at different times? Leaving aside the fact that that does not happen very often, it is a keen question because it does sometimes happen in very large applications of technology. But the question is on a par with asking a pianist, who has just explained the differing notations in the bass and treble clefs, how one can play with both hands at once and on the black keys as well as the white ones. The answer has to be: with further study and practice, including practice in mathematics, which cannot be made plausible in words beyond a certain point. As was mentioned in Chapter One, some things have to be taken on trust; and in this section the particular thing is that there are enough expert practitioners of the method to solve the problems. We do not all have to be able to play the piano with both hands, as long as we understand that a group of large co-ordinated projects is like Beethoven's 'Emperor Concerto'; it needs a good discount-method pianist for an adequate rendering. He or she will manage the double arpeggios. What we have to do, in our role of critics of the whole cost method, suspicious of its social consequences, is to make sure the right music is on the music rack.

The basic discount rate and the service life

The most important piece of information for any discount-method pianist, even a one-handed (present value or equivalent annual cost) performer, is the discount rate for a risk-free investment, which we can call the 'basic discount rate'. It corresponds to the musical key in which the piece is to be played. It depends on the financial arrangements for raising capital, which often involve a mixture of debt and equity. It is then an important – though often sadly neglected – task in making the best use of the cost method to distil the rate from the malt of those arrangements and base all the design decisions on that rate, with appropriate increases to allow for the different financial risks attaching to the various parts of the project and the various design options.

If that distillation runs into difficulties, as it often does in large organisations, it can sometimes be assisted. Generally a low discount rate favours options in which more capital is invested to save operating costs and vice versa. Hence the calculations can be repeated systematically over a range of possible rates to find the rate at which any two options have the same total present value or total equivalent annual cost. Thus in the generating station example, if we increase the discount rate to 20%, the present value of £100,000 a year for the next 25 years falls to just under £500,000. So, if the financial risk is judged to require a discount rate which is 5% above the basic rate, the question: 'What is the basic discount rate?' can be rephrased as: 'Is it more than 20 – 5 = 15% ?', which may be much easier to answer. If the answer is: 'Certainly not,' then the additional £500,000 can be invested without more ado. Or, if the £100,000 a year will diminish by 5% every year, the corresponding calculations lead to the question: 'Is the basic discount rate more or less than 14 – 5 = 9%?' (the calculation is explained in the appendix) and so on. But such calculations are really only a palliative. They cannot be repeated for all the hundreds of decisions in even a minor project, and they may be counter-productive if they become an excuse for postponing or neglecting the distillation of the basic discount rate for the project. It is an awkward but inescapable fact that the best design of any application of technology depends on how it will be financed. The problem is further complicated by monetary inflation and we shall therefore return to it in the next section.

Missed opportunities

Notwithstanding the superior merit of the discounting method over the payback-period method, natural conservatism and what are described euphemistically as 'internal communication problems' in large organisations will support the continuance of the latter method in many projects. Is that bound to reduce the contribution to prosperity of those projects? Could not the discount rate and the service life be used to calculate the cut-off period? Indeed they could; the formulae for doing so are almost the same as the formulae for the present value (see the Appendix 1 again). But there is no real merit in such a contrivance. It does not dispense with the necessity of distilling the basic discount rate and estimating the service life, and any organisation which can be persuaded to undertake those tasks and then to calculate the cut-off period, rather than just stipulate it arbitrarily or fix a budget out of the air, might as well go the whole way and adopt the discounting method in full. The payback-period method is potentially harmful; the actual harm comes from stipulating short cut-off periods, such as four years or less, arbitrarily.

Using the same formula once again, one can calculate the discount rate implied by any stipulated cut-off period. Thus for a service life of 25 years, a cut-off period of 11.7 years implies a discount rate of 7% a year. What rate does a cut-off period of four years imply? The answer is 25%, which is extremely high. But perhaps 25 years is an exceptionally long service life. Factories and chemical process plants are some of the places where the payback-period method is strongly established, and many of them are not planned to operate for more than ten years – though most of them go on for longer than that once they are built. So what discount rate does a cut-off period of four years imply if the service life is ten years? The answer is still 21% a year.

The payback-period method is sometimes commended for its supposed prudence, but it is not prudent to miss strings of opportunities to invest money at profit rates of up to 21% a year for ten years, which is what a cut-off period of four years implies in a project with a service life of ten years. 'Myopic' would be a more accurate description. Such an attitude would not matter so much if its adverse effects were restricted to the organisations in which it prevails, but they are not. The benefits of a well-designed project and the harm or loss of benefit of a poor design are felt in the community at locations far distant from the project itself, as we have already seen in Chapter Two.

Some of the biggest – and saddest – missed opportunities for deriving prosperity from technology occur when the payback-period method is used to design projects with very long service lives. It is not often used for generating stations (or at least it was not in Britain before the electricity supply industry was privatised) but it is for hospitals, which commonly stand for 100 years and seldom for less than 50 years, except when they are destroyed in wartime or by accident. Most of the costs of building and running a hospital are for accommodation, lighting, heating, ventilation, cooking, laundry and so on. Not all such services will last as long as the shell, but ten years is a conservative estimate for most of them. Yet the payback-period method is often used, with cut-off periods as short as four years – or even two years when financial conditions are supposedly stringent. It seems not to be realised that the method saddles the hospital with operating costs which could have been economically reduced by borrowing at interest rates in excess of 20%, with almost no risk of defaulting on the loan. The enormity of such losses is usually concealed because the operating costs and capital costs are rigorously separated in the accounts, without regard to the fact that part of the capital cost ought to have been incurred for the express purpose of reducing the operating costs and its success or failure in doing so ought to be assessed for the benefit of future projects.

The payback-period method is also sometimes commended because it avoids the necessity of estimating the service life. That is rather like commending the amputation of a limb because it avoids the necessity of treating a wound antiseptically. A live patient with only one leg is better than a dead patient with 'gangrene' on the death certificate, and a generating station which produces expensive electricity is better than one which runs out of finance before it is finished, but the arts of medicine and project design are the very arts of avoiding such stark choices, except in rare emergencies.

Suppose our best estimate of the service life is that it will be between 20 and 30 years, that is 25 years minus or plus 20%. Then with a discount rate of 12% a year, the present value of an annual saving of £100,000 lies between £747,000 and £806,000, that is £777,000 minus or plus 4%. Such a degree of uncertainty is not likely to affect the designer's choice appreciably, if at all, and it will be even smaller if the annual output diminishes from year to year. In sum, it is always better to face the difficulty and make even a poor estimate of the service life than to run away from the problem into the wooden arms of the payback-period method.

It turns out, ironically, that the length of the service life has more effect on the annual accounts than on the calculations in the cost method, and accountants might legitimately complain (though in fact they seldom do) that technologists generally give too little attention to improving the accuracy of the estimated service lives on which they, the accountants, have to rely. We shall return to that subject in Chapters Four and Five.

MONETARY INFLATION

Technologists recognise two kinds of monetary inflation, namely general inflation, and increases in specific costs. In Britain there was very little general inflation in peacetime until after the Second World War, and until then most people in Britain, including some economists, believed in the absoluteness of the unit of their currency. Some people still do and, judging by articles, tables and graphs in the financial press, they include some financial journalists and analysts, who even today often fail to make allowance for inflation in their presentations of long-term trends. But ordinary people, some of whom have felt at first hand the sensational consequences of inflation on their fixed pensions, have picked up some street wisdom over the last 50-odd years. We need to understand the origin of, and rationale underlying, that wisdom.

In Britain, electrical technologists talked about general inflation in the 1950s, when it was sometimes argued that hydro-electric generating stations, many of which were being built then in Scotland, were inherently more economical than coal-fired stations because, as large capital investments, the hydro stations provided a hedge against inflation and the water which spun their turbines was then free. But it was counter-argued from the other side of the meeting halls that oil wells, considered in conjunction with the thermal stations they supplied and the pipe lines between them, were also large investments and that the water for a hydro station was free only at its source, half way up a remote, high mountain, by which token oil was also free at its source, buried under the ground – or latterly under the sea. Such debates usually missed the relevance of the method of financing and so did not solve the problem, but they cleared the air to some extent.

Doubtless there were parallel debates in other institutions and countries, until the relativity of the value of money was gradually recognised by almost everyone, except the said financial analysts – and the Current Cost accountants, but we shall come to them in Chapter Five. The change in outlook did not and does not require anyone to abandon the well-established ways of assessing money values at any particular point in time. What it does demand is the adoption of a procedure for adjusting the money values of past times and future times whenever they have to be compared with those of the day.

The procedure, which is almost easier to perform than to describe in words, is to divide one of the values by the general price index at that time and then multiply the answer by the index at the time of the other value. So to compare the cost of a generating station which cost £200 million in July 1975, when the index was 138.5, with that of an exactly similar station in July 1985, when the index was 336.5, we calculate that (200/138.5) × 336.5 = 486 and say that the real cost at the July 1985 general price level of the station built in 1975 was £486 million. That does *not* tell us that such a station would have cost £486 million in July 1985. What it *does* tell us is that, if such a station actually cost £350 million in July 1975, the real price fell by 28% between those two dates because, although the money price rose by 75%, the value of money fell by 59%. That 28% fall is a piece of information which has been disentangled from the change in the value of money, and it is a useful, though small, piece of information to designers of generating stations, whereas the 75% rise by itself is just a piece of boring nonsense.

In Britain the general index is the Retail Prices Index, or RPI for short. It represents the inverse of the value of money when it is spent by the great majority of British households; it is thus essentially a consumer price index and the corresponding indices in other countries are often known by that name, which is a better name in as much as household spending is not confined to the retail market in any country. It might have been easier to teach successive rising generations about inflation if the index represented the value of money rather than its inverse, but there is nothing to be done about that now.

Some members of the accounting profession, and others, have proposed the use of special indices for particular goods and services used in particular industries, such as the 'construction industry', the 'engineering and allied industries other than vehicles' and so on, but such proposals go against one of the fundamental purposes of money, which is to provide the measure of value (utilitarian value of course) throughout an economy. Prices of individual things and restricted classes of things have always gone up and down, even when there has been no general inflation, but there is no need to doubt our instinctive feeling that a pound is a pound is a pound, *however* you spend it, though not *whenever* you spend it. One does not change one's outlook on an important issue without a compelling reason and it is only when prices in general go up and up so persistently and so much as to threaten the functioning of the whole economy that any fundamental change of outlook becomes necessary. Fortunately, after a long struggle, almost all the influential sections of society, at least in Britain, have adopted the RPI as the relevant index whenever inflation is under discussion[10].

But how do those calculations help in the design of future projects? Well their direct help is only slight, but the underlying assumption is fundamentally helpful. The assumption is that this kind of inflation affects all the costs equally. If we are designing a generating station and we expect that inflation will run at, say, 6% a year throughout its expected 25-year life, then in the absence of other information we can assume that its operating-cost prices – of its fuel, its staff and its stores – will all go up together at the same 6% annual rate. We shall return to the effects of other information, on the few occasions when there is any. Meanwhile it is not difficult to insert the assumption into the design calculations. Let us see what effect it has in the examples we have been considering.

When a general index becomes rather large, governments find it convenient to set up a new series in order to lessen the insistence with which the index reminds every user of the enormous degree of cumulative inflation in the long term. We have had five new RPI series in Britain since 1945 and two in the last twenty years. To simplify our arithmetic, let us assume a new series has just been set up at the start of our project; in other words the RPI is now 100.

Our example in the previous section was a set of design changes which will save £100,000 a year throughout the 25-year life of a generating station. Inflation at 6% a year will lift that saving, year by year, to £420,000 in the final year, but the procedure for adjusting the annual savings for comparison with the present cost of the proposal will bring them all down again to £100,000 a year at today's general price level. The next step is to apply the discounting formula, but before taking it we must consider the effect of inflation on the discount rate.

In the elementary example, if £100 is borrowed at 7% compound interest, the amount owing grows to £197 after ten years. But if inflation runs at 6% a year, the general index will rise from 100 now to 179.1 by then, so the adjusted amount which the lenders (using the cost method) will compare with today's £100 is £(197/179.1) × 100 = £110. In other words, at the end of ten years they will really be only £10 better off, at today's general price level, than when they started, not £97 better off. But that £10 represents less than 1% a year (0.94% to be more exact), which is to say that inflation at the rate of 6% a year reduces the effective or real rate of interest by just over 6% a year. Similarly if the money rate is 12%, the procedure reduces the amount owing after ten years from £311 to £173 at today's general price level, and the real rate of interest from 12% to 5.66%, that is again by something over 6%. The rule of thumb, which is widely known by now and good enough for most purposes, is to subtract the percentage rate of inflation from the percentage money rate of interest. The accurate formula, which is sometimes necessary, is explained in Appendix 2.

Thus a project financed by a loan at a fixed rate of interest for the duration of the project does indeed provide a hedge against inflation (though not of course a bigger hedge than another project of the same size, financed in the same way), in as much as the higher the rate of inflation the greater is the present value of the savings produced by increasing the investment. The calculations can be presented either in money or in adjusted values. In the former case, 6% inflation lifts the savings by 6% every year to £420,000 in the final year; they are then discounted at that fixed money rate plus a risk allowance, say 12% in all, to yield a present value of £1.32 million. In the latter case, the adjustment procedure brings all the savings back to £100,000 in every year but the real discount rate, as calculated from the money rate of 12%, is only 5.66%, which brings the present value up to precisely the same figure – £1.32 million, compared with only £780,000 with zero inflation. Similarly 8% inflation would increase the present value to £1.6 million, 10% to £2.0 million and so on.

The present value is always at today's general price level, however it has been calculated. So these increases in the present value are perfectly real; there is nothing imaginary about them. But of course if the lenders also expect inflation then, in the absence of other influences, they will increase the money rate of interest to restore the real rate. A 6% inflation rate will require a money rate of 18.7% to obtain a real rate of 12%, and 13.4% to obtain a real rate of 7%. (The rule of thumb is of course to add the expected rate of inflation to the intended real rate of interest.) So if the discount rate follows those increases it will pull the present values down to £1.17 million and £750,000 respectively, that is to the values which came out of the original calculations, assuming no inflation. It is therefore reasonable to presume that, if money can be borrowed for a period at a fixed rate of money interest, the rate includes the lenders' allowance for the expected rate of inflation, which can therefore be subtracted from the money rate when calculating the real discount rate.

No such adjustments would be necessary if projects were financed by 'index-linked' loans, that is to say loans in which the amount owing at the end of every year is first incremented by the rate of inflation during that year, then further incremented by the contracted real rate of interest and then reduced by any repayment at that time. Such loans would disentangle the cost method from the effects of inflation and so enable the projects to be better – often much better – designed. Until two decades or so ago, a suggestion like that would have been dismissed as totally impractical and thoroughly irresponsible, if not actually subversive, at least in Britain, but nowadays the British Government issues index-linked gilts without ruining the economy, although the outstanding index-linked debt probably puts a brake on tempting but inflationary public spending. Some changes in taxation legislation, which are considered in Chapter Five, would be necessary for the practice to become general, but there is no doubt that the deleterious effects of inflation on applications of technology could be neutralised if the benefits of doing so were fully understood.

Meanwhile projects are financed by mixtures of short-term loans, long-term loans at fixed or variable money rates of interest, debentures, preference shares, ordinary shares and so on. There are two options for setting up the design calculations and cost comparisons: to specify a money basic discount rate plus a corresponding estimate of the rate of inflation, and then make all the calculations in money values; or to specify a real basic discount rate and then make the calculations in adjusted values.

Technologists naturally prefer the second option because it involves them in less work and avoids the appearance of hugely inflated operating costs or savings in their calculations (a more than fourfold increase in our example), but it is then vital that the basic discount rate is the best estimate which can be made of the real rate, not the money rate. The effects of inflation are then concentrated on a single figure, which can be kept always in evidence, rather than allowed to percolate through the project and possibly be lost or distorted in the process.

If in a large organisation all the people responsible for and concerned with the basic discount rate genuinely accept the relativity of the value of money, the process of distilling that rate will go more smoothly, but they will not then be out of the woods because the rate of inflation does not remain constant, nor do interest rates go up and down nicely in phase with inflation. The reason is that governments try to reduce inflation largely by manipulating short-term interest rates, which then have complicated and often unpredictable knock-on effects on long-term interest rates and the other sources of finance. In such circumstances the problem becomes intractable and the contribution of technology to prosperity inevitably declines. For if the basic discount rate is pitched too high false economies are made, whilst if it is too low proposals with illusory savings are adopted. There is no safe way out. That decline is additional to the commercial and social harm which are commonly – and no doubt fairly – attributed to inflation and such attempts to control it.

Those are the difficulties of distilling the discount rate for a project which is already part of an investment programme and for which the method of financing has been decided or at least is under serious consideration. One might expect the difficulties to be even greater if a project has not yet reached that stage – if it has not yet been decided to invest at all in that kind of project. In the study of EVs in Western Europe, there were no serious plans for acquiring EVs in the numbers envisaged and so no possibility that the potential owners would themselves decide their basic discount rates.

However there is some evidence that, in the long run, the real rate of interest or profit actually obtained on safe loans or investments in Europe seldom exceeds about 3% a year, that is to say that higher money rates have been due to inflation or financial risk or both. The participants in the study therefore adopted a real basic discount rate of 3% and real rates, including the risk element, ranging from 5 to 10% in the cost comparisons. The particular comparison shown on page 72 was for a 5% real discount rate and yielded an advantage to the EV of 210 ECUs a year at the 1985 general price level in Sweden. At the other end of the range, a real rate of 10% reduced that advantage to 110 ECUs. A rate of 15.5% would have reduced it to zero. However some comparisons were more sensitive to the discount rate while others were less sensitive. More about sensitivities in the section on uncertainties.

Discount rates of 5% and 10% probably appear absurdly low, in the light of the money rates of interest at that time, to anyone who believes in the absolute value of money, but that sort of reaction has to be judged in relation to the change of outlook which such a person has failed to adopt[11]. An analogous change of outlook was required when it was discovered that the earth is a globe. One modern consequence of that discovery is that aircraft which fly from London to New York do not set off along a constant compass bearing towards New York, which is 73° West and 41° South of London. The aircrafts' initial bearing is 30° to the North of the constant bearing, which must appear absurd to anyone who believes the Earth is flat, but the Flat-Earther's journey would be 320 km longer than the shortest course, which starts off in that apparently wrong direction[12].

Few people would deny explicitly that the earth is a globe or nowadays that money values are relative, but it is not enough to assent to these propositions and then forget about them. They have to be thought right through and applied to actual flights and actual projects.

Increases in specific prices

The other kind of inflation recognised by technologists has a different nature. When general inflation is rapid, the cost of doing the work of a contract increases even during the period of the contract – more of course if that period is several years than if it is only a few weeks or months. Nowadays, with inflation at a low level, any such extra costs are generally borne by the contractor, but when inflation was serious they were normally passed on to the customer on the ground that they were not in the control of the contractor. However the prices of the various resources, notably materials and labour, which contractors require did not all rise together at the same rate; they moved irregularly and at different rates from one another and from the general price index. To enable the effects of such price increases to be passed on retrospectively without involving the contractor in additional profit or loss above or below the level included in the contract, a formula for adjusting the final payment was incorporated in the contract. The formula referred to one or more specific price indices, selected for their relevance to the type of contract. The Wholesale Price Index and indices of wage rates for various grades of labour were used in that way.

In as much as large proportions of the changes in these indices were undoubtedly caused by general inflation, it would be hypercritical to cavil at the practice of referring to them as a kind of inflation. One might try to bring them within the purview of the cost method by applying the adjustment procedure to them and labelling the adjusted indices as 'real specific indices'. But the real price changes indicated by the real indices would not be intrinsically more interesting than the real price changes of other goods and services for which specific indices have not been composed. Occasionally real price changes can be reliably forecast and the forecasts are of course pounced on by all the designers who can get their hands on them. Those are the rare occasions when one does not assume that all the money operating costs will go up together with general inflation; but specific indices – adjusted or not – of past prices are not particularly useful in that respect.

Are contractors really as helpless as they sometimes profess to be in controlling their costs? Is not the question whether the customer or the contractor can better predict future changes as relevant, or more relevant, to fixing a 'fair' price in some sense, than whether either party can control them? Do retrospective price adjustments weaken the contractors' grips on their suppliers and employees? Which party is better able to force the other to take risks which it is really in a better position to take itself? These are questions for dispassionate consideration in the circumstances of different types of contract at different times, but they are not directly relevant to our present pursuit of the cost method.

TWO KINDS OF COST

In Chapter Two, two legitimate meanings of 'cost' were mentioned, but there is no universally recognised pair of labels for distinguishing between them and they are often confused for that reason[13]. In this book 'commercial cost' will mean what the buyer or user pays or foregoes for something and 'resource cost' will mean what the community must forego in order to have that something, with the usual convention that the first word in either term can be omitted wherever the intended label is apparent from the context. Up to now in this chapter, under that convention, 'cost' has meant 'commercial cost'.

An additional source of confusion, which is not so easily eliminated, is that the meanings themselves are not totally unambiguous, as we shall see. But it will be contended that they are clear enough for our purpose (to derive prosperity from technology), provided we keep that purpose firmly in mind, to the exclusion of conflicting purposes, whenever we encounter ambiguities. Meanwhile, for knowledgeable readers, the commercial cost is not always the same as the purchase cost, because the relevant payments often do not involve a purchase (e.g. the cost of taking a bus to the station), nor always the same as the user cost, because the payments do not invariably involve the users (e.g. the cost of a new bus for the fleet, which may or may not be passed on to the passengers who travel in it).

The resource cost is not quite synonymous with what economists call the factor cost, in as much as the definition of 'factor cost' normally excludes the scarcity value of a resource. In the author's experience, the term 'opportunity cost' often disrupts the flow of a discussion among non-economists, which is a good enough reason for discarding it in favour of 'resource cost'. An additional reason is that the dictionary meaning of cost already includes the notion of something lost or foregone, which is sufficiently non-committal to include the notion of an opportunity foregone.

For economists and others who would still prefer 'opportunity cost' to 'resource cost', it is necessary to stipulate that the opportunity must be foregone by the community, not just the purchaser or user. The tax which is included in the commercial cost of the petrol for my car is part of my opportunity cost of my journeys in it, because I must forego other opportunities for spending that money in order to make those journeys, but that tax is not part of the resource cost (as defined) of those journeys because the community does not forego anything on account of those payments; they are balanced by equal and opposite gains to the Treasury. The essential point is that 'commercial cost' and 'resource cost' are both intended as labels, not as independent entities with a possible significance apart from their definitions.

The contribution of an application of technology to the prosperity of the community is greatest when its design is based on its resource costs, but except on rare occasions technologists base their calculations on the estimated commercial costs of their projects. Most of the time they do so automatically, without reflection, but if they wonder sometimes whether they ought to base them on the estimated resource costs instead and what would be the effects of doing so, their own financial interests or those of their employers are powerful inducements to continue in their normal habits – and there is no reason to blame them for that. The cost method is capable of distinguishing between the two kinds of cost but the distinction is seldom drawn. Consequently any differences between the two kinds of cost are potential threats to that prosperity. The actual loss of prosperity depends on the magnitudes and distribution of the differences.

Taxes

The most obvious – or least obscure – causes of such differences are taxes. One of the purposes of taxes is to raise government revenue and in doing so some taxes unfortunately raise the commercial costs of the taxed goods or services well above their resource costs, but at least one important revenue-raising tax does not have that effect, except at the consumer end of the economy where it does little if any damage. It is the well-known Value Added Tax, or VAT for short, and it was presumably designed with that objective in mind. Readers who are familiar with how VAT operates can skip the next two paragraphs without loss of continuity.

Although all trading organisations, except very small ones which are excused, have to register for VAT, they are in effect merely tax collectors. They add VAT to their invoices at the prescribed rates, which depend on their VAT categories, and (except when that rate is zero) they pass on the amounts thus collected to the tax authorities. But before doing so they deduct whatever VAT they have paid on the goods and services they have had from their suppliers at whatever rates applied to those suppliers. Thus their net VAT payments[14] depend on the value they have added to the goods and services they have supplied or provided; hence the name 'Value Added Tax'. But the only expenses to which they are put themselves are in the administration of the collections and the payments and in keeping the necessary accounts. Otherwise their costs are not affected by VAT, whatever the rate or rates, and their technologists can and do ignore it in designing their projects.

In the end the whole revenue from VAT comes from the ultimate consumers, who are not registered and have no customers and so no way of recovering the VAT which they have been charged. But apart from making may of them individually poorer (though not collectively poorer) than they otherwise would be, the only effects of VAT on the ultimate consumers are: (1) to bias their purchases away from goods and services on which the VAT rate is high towards those on which the rate is lower if the categories are to any extent interchangeable, and (2) to encourage them to engage in 'Do-It-Yourself' activities, for example by building their own garages – or even their own houses – thereby paying VAT on only the materials instead of on the whole operation. The first effect can be kept in check by charging VAT at the same rate on interchangeable goods and services, such as energy for heating premises and insulation materials. The second effect is unlikely to cause any serious diminution of prosperity because most applications of technology in a modern economy require more capital investment and a wider distribution of their goods and services than individuals or groups working without payment can muster.

The British Government makes rather heavy weather of administering VAT and causes some headaches among retailers, but that is mainly because it insists on having different rates of tax (zero, 5% and 17½% at the time of writing) on different goods, even when they are retailed in the same shops. In the eyes of a rational technologist, aside from his or her probable personal antipathy to paying any taxes, VAT is a brilliantly conceived tax.

Taxes on earned income (including in Britain national insurance contributions) raise the commercial cost of employing people, that is their gross income, above the resource cost. As long as there is full employment, the resource cost is their net income after deducting the tax[15]. The taxes therefore tend to bias the design of projects away from employing people towards more capital-intensive methods of production, but often the cost of a more capital-intensive method is ultimately composed largely of the costs of employing other people in various capacities, in which case income tax is not much less effective than VAT in raising revenue without excessively reducing the contribution of technology to prosperity. In either case it is not so much the tax itself as the charging of different rates on different, interchangeable goods and services or people which sometimes distorts the designs of projects and must therefore be justified by other criteria than the prosperity of the community as a whole. (When there is substantial unemployment, the resource cost of employing people may be appreciably less than their earned income, net of tax, but the subject of unemployment has a political flavour and it is therefore not considered until Chapter Eight.)

In the absence of inflation, analogous arguments apply to taxes on unearned income, which include interest on loans and profits from investments. But both depend on postponing the enjoyment of present wealth, which complicates the effect of those taxes when there is inflation. The problem and the solution will be easier to present after the two main problems in annual accounting, which were mentioned in Chapter One (page 10), have been tackled in Chapters Four and Five.

Other taxes create large and sometimes arbitrary differences between the two kinds of cost. Perhaps the most notorious in Britain in this century was the horsepower tax on passenger cars between the two world wars. The aim was to charge people who ran large cars with powerful engines more for their annual road licences than people with smaller cars. But some technical knowledge is required to measure the horsepower of an engine and it depends partly on how the engine is operated. To avoid having to run tests on the engines and to simplify the calculations, an arbitrary formula was adopted which had been devised seven years earlier and took no account of the scope for increasing the real horsepower of an engine by changing some features of its design. Further details are given in Appendix 3.

Consequently the tax induced the manufacturers in a competitive market to distort the designs of their engines in order to reduce the taxes paid by their customers while keeping up the real horsepowers. Those changes led to higher manufacturing costs, more maintenance, shorter engine life and other disadvantages – and also reduced the appeal of the cars in export markets. Happily the tax was replaced by a uniform tax after the Second World War, but it should never be forgotten as an example of the damage which uninformed politicians and their ignorant administrators can wreak on an important industry. Historians might draw a parallel with the Window Tax in the seventeenth century, which distorted the designs of houses in an analogous manner.

As already mentioned, the Excise duties on petrol and diesel fuel for road transport create differences between the commercial costs and the resource costs of those fuels, and they have had commensurate effects on the designs of motor vehicles[16], as witness the wide differences between European cars and their American counterparts in the heyday of the American motor industry. One purpose of those taxes is, of course, to raise revenue, but another in some countries is or has been to obtain a contribution, or increased contribution, from the owners or users of road vehicles to the public costs of providing and maintaining the roads[17]. Without such a contribution, too many roads would be provided or the demand for roads would outstrip the supply even more than it does now – with the taxes in place – in most Western countries. Fuel taxes are a crude way of trying to match the contribution of a vehicle or class of vehicles to the cost it imposes on the road system[18], but they are much better than nothing in that respect.

The European study of EVs was one of the rare occasions when it was possible and relevant to base the comparisons on resource costs rather than on commercial costs. If ever EVs proliferate in Europe, their designs will of course be governed by their commercial costs at that time, but their contribution to European prosperity will be governed by their resource costs. Those resource costs will certainly include the costs the vehicles impose on the European road systems but they are the same, or nearly the same, for EVs as for their conventional counterparts, so they could be omitted in the study. Thus the participants avoided the difficult task of calculating the cost which any particular vehicle or class of vehicles imposes (not that that task is impossible but it would have required more time, effort and insight than the participants could have provided). The *ad valorem* duties on the purchase costs of some vehicles and the costs of their annual road licences, both of which were different for different kinds of vehicles, were also omitted.

In that study, the immediate effect of removing those taxes from the cost comparisons was virtually to eliminate the economic advantages of the EVs, even when they would be produced in large numbers. The comparison (in ECUs) quoted on page 72 was then as follows:

	Electric	Petrol	Advantage of EV
For energy + maintenance	220	510	290
Equivalent annual cost of vehicle	1220	1250	30
Equivalent annual cost of battery	310	–	–310
Totals	1750	1760	10 (i.e. negligible)

That was not the end of the matter because up to that stage no account had been taken of the environmental advantages of the EVs (see page 53). However calculating those advantages, as distinct from just reciting them, broke new ground at the time and we shall come to those calculations in Chapter Nine.

Scarcity values

The prices of petrol and diesel oil in the comparisons of resource costs represented the price at which petroleum was imported into Europe, or at which it could be exported from the North Sea oil fields, plus the costs of refining the petroleum and distributing the two refined fuels. That petroleum price included the oil companies' costs of exploration and production, but it greatly exceeded their sum because petroleum was and is a scarce commodity in Europe. The difference arose from the substantial taxes levied by the governments which issued the drilling licences. The status of those taxes had to be resolved.

All taxes are transfer payments from the taxpayers to the tax authorities and, if one confines oneself to linguistic arguments, it is difficult to resist the conclusion that any tax which affects the commercial cost of a commodity must be excluded from the resource cost of that commodity. But every tonne of petroleum saved by a fleet of EVs could be either exported or consumed in place of a tonne of imported petroleum, at that price including those taxes; hence that was the relevant price in calculating the contribution, if any, of EVs to European prosperity.

In general every valuable resource, whether it be classed as land, labour (with full employment), a raw material or capital, has a scarcity value which must be included in its cost in design calculations. It is unusual for part of that scarcity value to be siphoned off as a tax, but the principle is sound, as most economists would agree, although they habitually prefer to present the argument in terms of 'opportunity costs'. The same principle can be adapted to calculate the environmental disadvantages – and sometimes advantages – of applications of technology, as will be described in Chapter Nine.

Subsidies

Subsidies, which are usually paid by governments but sometimes also by other bodies, make the commercial costs of the subsidised commodities or projects less than their resource costs. They thereby induce the designers of projects to use more of those commodities or invest more heavily in those projects than they otherwise would, and so they inevitably reduce the contributions of the affected projects to prosperity. Subsidies may be justified sometimes on other grounds, but the arguments – usually political arguments – in their favour would sound less hollow if their advocates published the amounts and displayed some understanding of their deleterious effects. In the European study of EVs, some governments were known to be subsidising the development and manufacture of EVs in various ways, and the manufacture of conventional vehicles as well, though to a less extent, but the subsidies were not eliminated from the cost comparisons because it would have been too difficult to elicit their true extent from those governments or the manufacturers. The conclusions of the study were weakened thereby, but our concern is with the method rather than the conclusions of any particular example of its application.

Profits

It is sometimes argued that profits raise the commercial costs of goods and services above their resource costs, but that ignores the element of risk in any project, particularly in the research and development phases. Perhaps that risk would be better understood if, whenever a company or undertaking produced and exploited a profitable innovation, such as a better battery, a cheaper way of making glass or a cure for a disease, it published a list of its failures over the previous 50 years and their costs. That is not likely to happen because it goes against the grain to publish one's failures, but somebody has to pay for them. Monopoly profits are a special case, which will be considered in Chapter Six, but willingness to take financial risks is a vital necessity for deriving prosperity from technology and its cost must be reckoned as a genuine resource cost.

Opponents of capitalism and the profit motive may argue that profits are normally much higher than the true costs of taking financial risks, even when the capitalists say they are competing with one another. But that type of argument can be dismissed unless the opponents can go on to explain how the true costs of taking risks can be estimated. That is just one part of the challenge in the last paragraph of Chapter Two.

UNCERTAINTIES

Nearly all the quantities in the previous sections – the commercial costs, the basic discount rate, the financial risks, the service lives, the rate of inflation, the scarcity values of resources and so on – have been presented as uncertain estimates, and whilst more co-operation and less secrecy could reduce some uncertainties, they will always be a major barrier to the theoretically best technology. Until computers, and particularly micro-computers, were invented, it was well beyond the skill of any designer or team of designers to juggle effectively with all the uncertainties in any project at once. It is still very difficult to do so but not quite so difficult as it was.

Part of the art (it is hardly a science yet) is to discover to which uncertainties the decisions are sensitive and to which insensitive. The technique is known, appropriately, as 'sensitivity analysis'. The repeated calculations of the present value on page 74, to discover whether the discount rate was higher or lower than the rate which equated the present value to the present cost, were an example.

In the study of EVs in Europe, the estimates were as uncertain as in most comparisons and much more uncertain than in some, so the calculations were laid out to facilitate sensitivity analysis from the outset[19]. Some of the more interesting results were:

1) the environmental advantages (when they came to be calculated) were crucial – without them the equivalent annual resource costs of the EVs were nearly all as great as or greater than the costs of their conventional counterparts on the same missions, even when the EVs were to be mass-produced;
2) the vehicle design and cost were also crucial – some EVs were hopelessly uneconomic (had much higher equivalent annual costs than their conventional counterparts) under most conditions, even when their environmental advantages were included, while others were economic under many different conditions;
3) some countries, notably the Nordic countries, were much more favourable to EVs than others, but they have small populations and so could not provide large enough markets on their own;
4) the discount rate, the daily mileages of the EVs (within their battery capacities) and their lower maintenance costs were less important than their longer lives.

Those results may not sound very surprising with the benefit of hindsight, but they were not obvious to the participants in the early stages of the study. They provided a basis of some confidence in the conclusions, one of which was that six million small electric cars and one million small electric vans could contribute roughly one billion ECUs (£600 million) a year at the 1985 general price level to the prosperity of Western Europe, almost wholly in the form of environmental advantages. That is not to say any great confidence was felt in the figure of one billion itself, but it would have required an implausible combination of circumstances to reduce the contribution to zero or a negligible amount and it might turn out to be an under-estimate. The recommendations for various actions flowed from the sensitivity analyses as much as from the cost comparisons themselves.

EXECUTION AND COMPLETION

Up to now the impression may have been given that technologists spend all their time sitting in offices, calculating and writing specifications. Certainly they spend much of their time in those ways and their low public profile, often amounting to complete invisibility when they are doing so, is one of the reasons why their functions are so poorly understood by the general public. They lack the advantage enjoyed by the medical, legal, military, ecclesiastical and teaching professions of being often in view, or at least in frequent contact with members of the non-professional public.

However the efforts of technologists to contribute to prosperity while sitting in offices come to little or nothing if their decisions and recommendations are not carried out meticulously on the factory floors, at the construction sites and wherever else they are supposed to be translated into reality. And notwithstanding the machines and computers which are employed wherever and whenever practicable, they are carried out in the last resort by men and women.

Some readers may be thinking that it is high time that human beings were included in the portrait of Oliver Cost-Method and perhaps that they should have been the first of his six aspects to be painted, not the last. In reply to that point, it is often an advantage to put the most important item at the end, particularly when, as in this case, it could not be at the beginning because one cannot execute and complete a project until one has obtained the estimates.

The central problem is how to get the people taking part in a project to co-operate and trust one another sufficiently to complete the project successfully. It comes under the heading of management and it used to be the boast of many top managers in large British organisations not many years ago that management is a largely self-sufficient field of expertise and that top managers need not know much about the products and services of their organisations. It is a dangerous belief and very frustrating to the middle management and work force. In the author's opinion, it is one of the causes of Britain's relative industrial decline in the twentieth century, coupled as it has been with a disdainful lack of knowledge of technology among many managers and non-managers alike. Fortunately both the belief and the disdain appear to be dying at last in Britain, though they are still far from dead.

Some organisations, particularly those engaged in the building of large engineering projects, accept that technologists have their part to play in internal management – the actual design work and the construction on site – but exclude them from the procurement of plant and equipment from external sources or minimise their participation in that part of the projects. The practice is then to require the designers to specify precisely what they require and normally (though not invariably) to allow them to inspect what is delivered or provided before it is paid for, but to disregard their preference for one supplier or contractor over another, unless it is based on lower quoted prices or other solid evidence which can be demonstrated to the satisfaction of the non-technologists in charge of external procurement.

Designers who have had previous contacts with several suppliers may say they prefer one rather than another for such apparently vague (to non-technologists) reasons as: 'They always do a good job' or, 'They are reliable' or, 'They won't sting us as hard or be so awkward as the cheaper people if changes are necessary or there is a dispute.' Such reasons are not ideal but they are often inevitable and should not be disregarded for two reasons.

The first is that writing specifications is not an exact science; they are sometimes ambiguous and, more to the point, they can often be twisted in a court of law to mean something which neither party's technologists genuinely understood before the dispute arose. For that reason technologists generally prefer to avoid litigation if they can and to deal with suppliers and contractors whose managements do not gag their own technologists in the event of a dispute.

The other, separate reason was alluded to in the section of this chapter on 'Obtaining the Estimates' (page 63). If, in the absence of any such ambiguity, the performance of a piece of equipment is worse (or better) than was specified, the long-term loss (or gain) to the purchaser is often far greater than can reasonably be specified as a penalty (or bonus) in the contract. Under that heading total rejection with no payment may still leave the purchaser with a substantial, unrecoverable loss.

And that applies not just to large items, such as electricity generators, but to an ever-increasing range of items as technology becomes more complicated and more reliance is placed on the correct operation of every part of a large project over its whole service life, which may be 20 or 30 years. The most dramatically appalling losses nowadays, sometimes involving human lives, arise from software errors in computer programmes which have been written under pressure to complete a project in time at a competitive price. They sometimes do not surface for years after the programme has been accepted and they may be almost impossible to rectify.

In such cases the lowest tender may not represent the best buy, even when it complies with the letter of the specification and the contractor's financial resources have been fully demonstrated. The practice, which is not uncommon in large international contracts, of requiring the successful tenderer to put up a 'contract bond', under which a bank, employed by the contractor, guarantees to pay the purchaser a penalty if the contract is not fully and correctly completed, cannot guarantee the best outcome in the long run if the penalty does not fully compensate the purchaser or if another tenderer would have provided goods or services of a higher quality than was originally specified.

The logical extension of that practice is to require every tenderer to put up a 'bid bond', under which a penalty is paid if a tenderer whose tender is formally accepted fails to put up the specified contract bond. The intention is to eliminate tenderers who hope to obtain contracts by tempting the purchaser to dispense with the contract bond rather than accept the next lowest tender at a higher (sometimes much higher) price, but bid bonds from all the tenderers inevitably increase the total cost of tendering, which the purchasers between them must pay in the end; otherwise unsuccessful tenderers will be driven out of business until those which remain are so few that they can successfully refuse to bid for contracts which require contract bonds. There is a practical limit to the effectiveness of legal or legalistic attempts to guarantee that any contract will be completed to the satisfaction of the purchaser. There has to be a judicious admixture of legal guarantees, judgement and trust.

Involving the designers of a project in all the stages of its execution and completion and deferring to their judgements, even when they are subjective, generally (though not invariably) helps to build up a code of what is known as 'business morality', in which the parties to contracts have regard for each other's financial interests as well as their own. The purchasers do not apply pressure and withhold payments whenever they think they can gain some advantage in the short term, preferring instead to maintain friendly relations, in the hope and expectation of further business which will be profitable to both parties in the future.

Business morality of that kind is presumably also practised by the parties in non-technological contracts. In technological contracts the technologists on both sides do not seek to exploit every weakness in the specification, but prefer instead to build up standard specifications and conditions of contract which are as unambiguous as possible, fair to both parties and at the same time reasonably easy to read and understand. The work of the leading 'engineering' (i.e. technological) institutions in that field has been particularly valuable.

REFERENCES AND FOOTNOTES

1) Walpole's *Anecdotes of Painting.*
2) COST 302 – 'Technical and economic conditions for the use of electric road vehicles', published by the European Commission, Brussels, 1987. Ten nations participated. The author represented Britain. ('COST' is an acronym, nothing to do with cost.)
3) In Britain the privatisation of the electricity supply industry in 1990 has separated the ownership of the generating stations from the ownership of the grid, but in some countries they are under common, not necessarily national, ownership. The British situation is considered in Chapter Seven.
4) One reason for the reluctance of EV manufacturers to estimate the costs of EVs and batteries in large numbers for the European study referred to on page 54.
5) The other reason why the EV manufacturers were reluctant to divulge the costs of EVs for the European study, which of course published the information. Their eventual willingness to do so was remarkable.
6) Sometimes referred to, ambiguously and confusingly, as the 'payback period' without regard to the difference between what is required and what is expected.
7) The discount rate is often referred to as the 'required rate of return' on the investment. That is a confusing description unless you are already familiar with it because it seems to mean the rate at which the investment must return or be paid back. In fact the 'required gross rate of return' meant just that in a Government white paper in 1967 (Cmnd 3437), and the 'required net rate of return' meant the discount rate. The dual terms were a source of confusion until the gross rate passed out of the vernacular and the 'net' was omitted from the net rate. In this section, the word 'return' is not used because it seems to hark back to the payback-period method, but it is now so entrenched in normal parlance that it would be futile to insist on excluding it in some contexts.
8) Strictly speaking, the present value of the final disposal or scrap value at the end of its service life should be subtracted. But with few exceptions the final disposal value is too small and uncertain to be worth bringing into the calculation.
9) One ECU was SEK 6.66 or 59 pence at that time.
10) The exception is the government, which usually prefers the so-called 'underlying rate', excluding interest on house mortgage

payments, but the difference is not serious in the present context because the two rates tend to converge over any period of several years.

11) Some technologists, who read the full report, felt that the risk allowances of between 2% and 7% were too little, but that was a separate criticism.
12) It swings round steadily towards the south as the great circle is traversed.
13) *Chambers English Dictionary* defines 'cost' as 'what is or would have to be laid out or suffered or lost to obtain anything', which is much more helpful than the circular definitions of 'prosper' and 'thrive', but it does not say who is doing or would have to do the laying out or suffering or losing.
14) Or rebates, if the deductions exceed the amounts collected.
15) The commercial cost also includes other items, such as pension contributions and accommodation, but they are also part of the resource cost.
16) Haphazard but probably beneficial in that case, in view of the adverse effects of the exhaust fumes on the environment. That subject is addressed in Chapter Nine.
17) See 'Report on Road Track Costs', Ministry of Transport/HMSO, 1968 and 'Rational Road Pricing Costs in Canada', Canadian Transport Commission Report, 1973.
18) Interestingly, in Sweden at the time of the European study of EVs, the contribution from vehicles with diesel engines was obtained, not by a tax on their fuel but by a tax on their annual mileages, as measured by their tachographs, at rates per kilometre depending on their gross weights, which was a much better way of matching the contributions to the costs imposed.
19) Serious users of computers will have heard of spreadsheet packages, which are ideal for this purpose. They enable the user to vary the items of data input, individually or in any combination, and repeat the entire set of calculations in a few seconds.

APPENDIX 1 – DISCOUNTING FORMULAE

Present value

£1 lent at annually compounded interest for m years will of course amount to £$(1+r)^m$, where r is the rate of interest, with the convention that '%' means '/100', so that 7% = 7/100 = .07. Hence the present value of £1 to be paid or received m years hence is the reciprocal of that amount, i.e. £$(1+r)^{-m}$, where r is now the discount rate. It follows that the present value, PV, of a constant stream of costs or revenues paid or received annually in arrears at the rate of A per year for n years is given by the geometric progression:

$$PV = A.\{(1+r)^{-1} + (1+r)^{-2} + \ldots + (1+r)^{-m} + \ldots + (1+r)^{-(n-1)} + (1+r)^{-n}\}$$

Multiplying both sides by $(1+r)$:

$$PV.(1+r) = A.\{1 + (1+r)^{-1} + \ldots (1+r)^{-m} + \ldots + (1+r)^{-(n-1)}\}$$

Whence, by subtraction and division by r:

$$PV = A.\{1 - (1+r)^{-n}\}/r$$

For a stream which starts at the rate of A per year but then immediately begins to diminish at the rate of s per year:

$$PV = A.\{(1+r)^{-1}.(1-s) + (1+r)^{-2}.(1-s)^2 + \ldots + (1+r)^{-m}.(1-s)^m + \ldots + (1+r)^{-n}.(1-s)^n\}$$

Whence, by multiplication by $(1+r)/(1-s)$, subtraction and division by $(r+s)/(1-s)$:

$$PV = A.[1 - \{(1+r)/(1-s)\}^{-n}].(1-s)/(r+s)$$

Both formulae slightly under-estimate the present value of a stream of savings by assuming that the operating costs are all paid annually in arrears, whereas in practice they are paid at different intervals (weekly, monthly or quarterly), but the error is too small to worry about in most cases.

A possible reason for the unpopularity of the discounting method up to about 1970 is that these formulae cannot be evaluated with a slide rule and are very laborious to evaluate with tables, mainly due to the double interpolations between the tabulated values of n and r, but modern hand-held calculators have eliminated the need for such labours.

From the second formula, if A = £100,000 and s = 5%, then, by trial and error, PV = £500,000 if r = 14%.

Equivalent annual cost

Inverting the first formula for the present value yields the following formula for the equivalent annual cost, EAC, spread over n years, of a capital expenditure, C, at the commencement:

$$EAC = r.C/\{1 - (1+r)^{-n}\}$$

Equivalent cut-off period

The equivalent cut-off period, ECP, can be calculated from the equation:

$$ECP \times EAC = C$$

whence

$$ECP = C/EAC = \{1 - (1+r)^{-n}\}/r$$

From that formula, if n = 25 years and r = 7%, then ECP = 11.7 years.

And if n = 25 years, then by trial and error, ECP = 4 years if r = 25%.

Finally, if n = 10 years, then by trial and error, ECP = 4 years if r = 21%.

APPENDIX 2 – MONETARY INFLATION FORMULAE

The procedure for adjusting the amount owing after m years is equivalent to dividing it by $(1+i)^m$, giving $\{(1+r)/(1+i)\}^m$, where i is the annual rate of inflation. The effective or real rate of interest, i.e. the rate which will yield that adjusted amount after m years, is therefore given by the equation:

$$(1+q)^m = \{(1+r)/(1+i)\}^m$$

where q is that real rate. Whence

$$q = (r - i)/(1+i) \approx r - i \quad \text{if } i \text{ is small.}$$

As far as the author is aware, both the formula and the approximation first appeared – using a slightly different notation – in his paper, 'A General Theory of Depreciation of Engineering Plant', published by the Institution of Electrical Engineers in November 1960 and printed in Part A of the *Journal* of the Institution in October 1961. However the subject was much discussed and the formula may have been proposed elsewhere at an earlier date.

The multiplier for the presentation in money values is

$$[1 - \{(1+r)/(1+i)\}^{-n}] \times (1+i)/(r-i)$$

and for the presentation in adjusted values it is

$$[1 - (1+q)^{-n}]/q$$

which is the same because

$$1+q = (1+r)/(1+i) \qquad \text{and} \qquad q = (r-i)/(1+i)$$

APPENDIX 3 – THE PRE-WAR HORSEPOWER TAX ON PASSENGER CARS

The horsepower developed by an internal combustion engine depends on the number of cylinders, the diameter and 'stroke' of the pistons (i.e. the distance they move up and down in the cylinders), the pressure of the gases in the cylinders and the speed of rotation of the 'crankshaft' (which is driven round by the pistons and in turn propels the car via the clutch and gearbox).

But the formula for the tax included only the number of cylinders and their diameter, not the piston stroke, gas pressure or engine speed. It had been devised in 1906 by the Automobile Club, which later became the Royal Automobile Club, and it was based on an assumed piston stroke, assumed gas pressure and assumed engine speed, taken from an engine designed at that time. The horsepower tax was not introduced until 1913.

Consequently the real horsepower could be increased by manufacturing engines with long strokes and running them at high speeds, while reducing their nominal horsepower by reducing their cylinder diameters. But the long strokes and high speeds increased the cylinder wear and the costs of those engines, compared with the optimum design (using the cost method) in the absence of that arbitrary tax.

4

TECHNOLOGY AND ANNUAL ACCOUNTING

> Annual income twenty pounds, annual expenditure nineteen nineteen six, result happiness. Annual income twenty pounds, annual expenditure twenty pounds ought and six, result misery.[1]

For many people (including the author) who have struggled at some time with an inadequate or barely adequate income to support themselves and their families without luxuries, this quotation may seem to epitomise the nature of accounting – a plodding (some would say sordid) preoccupation with the cost of material necessities. Looked at more objectively, accounting is not concerned only or mainly with poverty; its principles are – or should be – the same for rich and poor alike, but even in a large prosperous company annual accounting is essentially a pedestrian activity, as are routine physical measurements. However that does not mean that these activities have no important consequences.

The simplest form of annual accounting is exemplified by something like an ordinary amateur badminton club. The income comprises the members' subscriptions, plus perhaps a small profit on the refreshments served at the annual general meeting and the sale of some raffle tickets, while the expenditure consists of the hire of some courts in a hall and the purchase of shuttlecocks. As long as the income slightly exceeds the expenditure, the cash in hand will slowly increase and the honorary treasurer will be content, but if the expenditure exceeds the income the cash in hand will diminish and the club will eventually be unable to continue unless some redressing action is taken.

Badminton, like many leisure pursuits in prosperous communities nowadays, has benefited from modern technology. The rackets are made of high-tensile steel or carbon fibres and for the ordinary amateur game the plastic shuttlecocks are probably made in computer-controlled factories, while the badminton halls are of reinforced concrete, with anti-dazzle internal lighting, special resilient floors to reduce accidental injuries and sometimes air conditioning. But none of those technologies has any bearing on the accounting method of an ordinary badminton club because it does not own the hall nor the means of producing the rackets or shuttlecocks. The members can therefore continue, expand, contract or discontinue their activities without any notable accounting problems. If they wind the club up nobody will worry about a few unused shuttlecocks. They do not need a balance sheet to show the values of the club's assets and liabilities at the end of every year, nor the services of an accountant.

At the other end of the accounting spectrum, the annual accounts of a large company or undertaking are very complicated, for a variety of reasons. They must make provision for debtors and creditors, comply with a large volume of complex legislation, be proof against fraud and embezzlement, deal with fluctuating share prices of recently acquired minority holdings and so on and so on. But the only complication which concerns technology arises from the ownership of the plant, machinery, equipment and buildings by or in which the technology is applied, and the mines, oil wells and quarries, etc. which yield the raw materials and fuels for the applications. They are known collectively as 'tangible fixed assets' to distinguish them from other kinds of asset although they are not necessarily fixed physically, and for brevity they are often referred to simply as 'fixed assets'[2]. Fixed assets in that sense are at the centre of the accounting problem as far as technology is concerned, irrespective of the size or complexity of the owners' financial affairs. In the aggregate, they form the greater part of a modern society's accumulated wealth. How they are valued and how their depreciation (or fall in value) from year to year is calculated affect their contribution to prosperity.

THE SPECIAL PROBLEM OF ANNUAL DEPRECIATION

A traditional principle in annual accounting has been the principle of Historical Cost. Historical Cost accounts aim to represent only what is known to have occurred in the past, not what anybody believes will occur in the future. By this principle accountants strive to make accounting a reliable and objective science, like physical metrology. Or at least they used to do so until about 1979, when Current Cost Accounting was introduced in response to some bouts of severe monetary inflation in several countries, but inflation accounting (as it is called) is the subject of the next chapter. In this chapter we assume there is no inflation.

The Historical Cost principle works very well for recording the incoming and outgoing payments during the year of account and for the outstanding debtors and creditors at the end of the year – in fact for almost all the items in the annual accounts. But it has to be partially relinquished when it comes to the depreciation of the fixed assets because that depends on the estimated net value of those assets at the end of the year, which depends in turn on some estimates or assumptions about how they will be employed in the remaining years of their service lives. In other words last year's profit or loss inevitable depends on next year's plans – and to a decreasing extent on plans or assumptions for the more distant future. That is the special problem of depreciation.

The same is true of so-called 'current assets', which comprise the work still in progress at the end of the year together with the stores which are due to be consumed or sold in a subsequent year. But because the current assets (unlike the fixed assets) are intended for physical consumption or sale, the problem of valuing them is trivial in comparison with that of valuing the fixed assets.

For a fixed asset, the problem is normally divided into two parts. One part is how to estimate the useful life of the asset, that is to say the period after which it would be better in some sense to dispose of it than to continue using it. That is not necessarily the same as the actual service life because fixed assets are commonly kept in service long after it would be better – sometimes much better – to dispose of them; also they are sometimes disposed of prematurely. In that context, 'better' usually means cheaper and 'disposed of' usually means scrapped, but cheapness is not always the criterion and old fixed assets are sometimes sold for further use by others, particularly when the assets are transportable.

The other part of the problem is how to calculate the annual depreciation – in effect to decide how to spread the cost of the asset (less its final disposal value) over its expected useful life. Accountants normally opt out of estimating the useful life of an asset, either by leaving it to the technologists or more often by using conventional 'standard' lives for different types of asset, such as buildings, machinery, vehicles, etc. But they insist on calculating the annual depreciation themselves and commonly ignore the advice of technologists, even when they have taken the advice of the same technologists on the useful life of the asset.

How then do accountants calculate annual depreciation? A detailed answer is given in Appendix 1 to this chapter. What it amounts to is that, except for the special case of mines, oil wells and quarries, etc., which are known as 'wasting' assets, they insist that there are no objective rules which can be applied to the calculations.

They claim that, for non-wasting assets, their only firm data are the acquisition costs and final disposal values of the assets. Between those two limits the present Accounting Standard (see the appendix) requires no more than that the accountants shall disclose the method they have used, the effect of any change in the method and the reason for the change. Otherwise that Standard gives them *carte blanche*. Whichever method they choose, provided they use it faithfully, the auditors will use the same method and so confirm their results. The reasons why any particular method was chosen in the first place need not be – and generally are not – disclosed.

In practice accountants almost invariable choose one or other of two conventional methods. The first – and by far the commonest – is the so-called 'straight-line method'. It assumes that the annual depreciation is constant over the useful life and so can be calculated by simply dividing the acquisition cost, minus the final disposal value (if any), by the useful life. The key word is 'assumes'. The method does not rest on any knowledge or understanding of how fixed assets actually depreciate.

The other, less commonly used, conventional method is the 'reducing-balance' method, by which the annual depreciation is calculated as a fixed percentage of the net value at the beginning of the year, that is after deducting the depreciation in the previous years. Here again the fixed percentage is assumed. It does not rest on any knowledge or understanding of depreciation. But technologists have a large body of such knowledge and understanding, which deserves to be brought into the calculations.

Appendix 1 also answers the contention, which is often made, that, in a large company with a more or less continuous programme of investment in fixed assets, it does not matter much which method is used because they all tend to produce much the same answer. In fact they only exhibit that tendency when the company is neither expanding nor contracting and when the technology which is being applied is not changing rapidly. Under normal, non-static conditions, the method affects the annual profit or loss and/or the prices which even a large company must charge in order to make a profit or avoid making a loss.

It follows that, if the annual depreciation can be objectively estimated, then companies and undertakings which do not estimate it are either over-optimistic or over-pessimistic in some degree. In the former case they will be inclined to lower their prices too far and take too many risks when they are expanding, and their financial difficulties when eventually they must reduce their rate of investment will be exacerbated thereby. Moreover the whole community will lose by using and relying on their goods or services, which are in reality uneconomic and may be withdrawn precipitately when those difficulties are encountered. In the latter case, the companies will be inclined to raise their prices too high and take too few risks, and so perhaps fail to produce goods or provide services for which there would be a demand at a lower price and which would add to their prosperity and that of the community in which they operate.

THE CAUSES OF DEPRECIATION

The reason why accountants do not attempt to measure the annual depreciation of non-wasting fixed assets or estimate it objectively seems to be that, by and large, their leading researchers and contributors to accounting literature, having wrestled with the problem hard and long[3], have concluded regretfully that the absence of any commercial transactions between the acquisition and disposal of such assets precludes the possibility of ever solving it. But in reaching that conclusion, those researchers have repeatedly failed to perceive the difference – long recognised and extensively exploited in physical science – between direct measurements and measurements or estimates which depend for their validity on theory.

Such things as the number of apples in a parcel of groceries or the weight of a joint of beef can be counted or measured directly, but most of the knowledge which makes up the body of physical science has not been acquired by such means. In the words of a world-famous astronomer and physicist in 1939:

> Every item of physical knowledge ... is an assertion of what has been or what would be the result of carrying out a specified observational procedure. Generally it is an assertion of what *would be* the result if an observation were made; for this reason it is more accurate to describe

> physical knowledge as *hypothetico-observational* [that is knowledge of the result of a hypothetical observation]. ... I am not denying the importance of actual observation as a source of knowledge; but as a constituent of scientific knowledge it is almost negligible...
>
> Consider, for example, our knowledge that the distance of the moon is about 240,000 miles. ... Accurately enough for present purposes, what we claim to know is that 240,000 × 1,760 yard-sticks placed end to end would reach from here to the moon. That is hypothetico-observational knowledge; for certainly no-one has carried out the experiment. ... There are a variety of practical methods of finding the distance; one of the most accurate [at that time] involves *inter alia* swinging pendulums on different latitudes on the earth. Although it would be true to assert that 240,000 miles is the result of an actual observational procedure, that is not what we intend to assert when we say that the distance of the moon is 240,000 miles. By employing accepted theory we have been able to substitute for the actual observational procedure a hypothetical observational procedure which would yield the same result if it were carried out.[4]

Nowadays the distance is measured more accurately by bouncing electromagnetic waves off a reflector placed on the moon, but that has not changed the definition of the distance, which refers to a physically impossible measurement, nor the theory by which the observations of the swinging pendulums were converted into miles. Indeed the astronauts who placed the reflector and the technologists who designed their spacecraft relied on that hypothetical definition and that theory quite as much as on those observations in order to accomplish their mission. It is no less legitimate to rely on theory to measure or estimate annual depreciation indirectly.

Technologists have long been familiar with the causes of depreciation, namely physical deterioration, technological progress and diminution of demand – and how they operate –

and that familiarity can be utilised to make better calculations of annual depreciation than come out of either of the conventional methods, better that is in the sense that decisions based on those calculations would generally contribute more to our prosperity than decisions based on those methods. It is a case of acquiring knowledge in one field and applying it in a different field.

Take physical deterioration first. Some fixed assets, with individually small acquisition costs and short service lives, wear out physically, in the sense that physical changes in them eventually make them no longer usable for the purpose for which they were produced. Traction batteries for electric vehicles are a good example; their useful lives depend on how intensively and how carefully they are used but there comes a time when they can no longer store enough energy to give the vehicles a useful range before they have to be recharged. That time can be predicted from the design and construction of the batteries and reputable manufacturers used to guarantee it, with a pro-rata money-back refund if it was not achieved. If a battery is used to the same extent every year, the straight-line method, though not theoretically perfect, normally provides an adequately accurate estimate of its annual depreciation. But non-wasting assets which wear out physically constitute only a tiny proportion of any community's investment in fixed assets at any one time.

The overwhelming majority of non-wasting fixed assets do not wear out in that manner. The occasions when they become no longer physically usable for the purpose for which they have been produced are accidental exceptions, which are almost as likely to happen when they are new as when they are old and which are, or should be, covered by insurance rather than by depreciation provision. Normally, as an asset gets older, its annual operating cost per unit of output or utilisation gradually increases. For example, an electricity generating station gradually consumes more fuel and requires more maintenance per kilowatt-hour of output; similarly a passenger bus gradually requires more fuel and maintenance per kilometre of distance driven.

If physical deterioration were the only cause of depreciation, that trend would continue until a discounting calculation showed that it would be slightly cheaper to build a new generating station or buy a new bus than to continue running the old one. That result would mark the end of what is best described as the 'economic' life, to emphasise that it is not a physical event but the outcome of a cost calculation.

However, except for the aforesaid tiny minority of assets, physical deterioration is not generally the only cause or the main cause of depreciation. Careful maintenance can keep a generating station or a bus or any other such fixed asset in such a good condition that its fuel consumption and its need for maintenance per unit of output or utilisation hardly increase from year to year, in which case the asset hardly depreciates because of physical deterioration and its economic life would be very long if it were not for the other causes. That is true even of private motor cars. They are affected by use, exposure to the weather and so on, but if motorists knew with certainty that their next new cars would be exactly like their old ones when they were new and would cost exactly the same (remember, no inflation), they would want to postpone the renewals far longer than they do at present and cars would generally be much more carefully maintained and would become much older before going on to the scrap heap. In practice, fixed assets are less well maintained for the sound reason that it is uneconomic to guard against physical deterioration beyond the end of the economic life as determined by the other causes. So the assets do deteriorate, but that still does not mean that deterioration is the prime cause of depreciation.

Physical deterioration is internal; the other two causes are external. After a few years the design and performance of a new bus, which one would buy if the time were ripe, become better than those of the bus in service. A new bus may be cheaper or more fuel-efficient or burn cheaper fuel or have a better performance or require less maintenance or possess other qualities, including qualities (such as comfort and appearance) whose appeal is partly subjective. In general it will have a combination

Technological progress

of qualities (not necessarily including cheaper first cost because the other improvements are often worth paying more for) which the transport company would prefer.

A gradational process

That is the meaning of technological progress. It is inherently gradational rather than revolutionary. For example, in coal-fired generating stations built in Britain since the 1920s, steam temperatures have been increased in modest steps from 450°C or less to 565°C or more; the number of stages of bled-steam boiler feedwater heating has been increased (not all at once) from three to eight and the main flow of steam is now diverted back to the boiler for reheating twice during its passage through the turbine (in an intermediate design it was diverted once); the coal is pulverised instead of burned on moving grates; the generator conductors have been cooled with hydrogen instead of air and then with water instead of hydrogen; the thin steel plates which confine and conduct the magnetic fields in the transformers are now rolled cold instead of hot. All those developments and many others have improved the efficiency and reduced the operating costs of the stations, but they have done so in small steps because technologists are always aware of the risks of making changes, and especially of making several changes at once lest they should interact unexpectedly. The individual gains are usually modest but the losses if something goes wrong may be very heavy. The cost of losing the output of a modern generator is reckoned in thousands of pounds an hour.

Over the same historical period, ships have become very much larger and so cheaper per tonne of cargo capacity; their hulls are welded automatically instead of being riveted by hand; their superstructures may be of light alloys instead of steel, which improves their stability in rough seas, and they are navigated with the aid of satellites which can pinpoint their positions with great accuracy. An almost endless list of similar gradational improvements could be quoted from other branches of technology.

Sometimes technological progress is foreshadowed by a comparatively sudden scientific breakthrough, the most obvious example being the discovery of artificially induced nuclear fission during the Second World War. Yet the peace-time benefits in the way of cheaper electricity in the ensuing half-century have been gradual.

In some branches of technology, notably computer technology, the progress has been much more rapid. An ordinary home computer, costing a few hundred pounds in 1996, has more computing power than the largest computer in Britain, costing several million pounds, 25 years earlier. But even that progress has been essentially gradational. In general we can speak of 'the rate of technological progress' in an industry or branch of industry, which is analogous to the slope of an irregular path. We can often gauge the slope or rate in advance, but not the individual irregularities nor just when they will occur.

The other external cause of depreciation is the diminution, or prospective diminution, of the demand for the good or service produced or provided by the asset. Sometimes the utilisation of an asset is deliberately reduced by the owner in order to favour other, generally newer assets with lower operating costs, notably in electricity generation as mentioned in Chapter Three. In other cases, the reduction is caused by the actions of a competitor, such as a rival generating company or bus operator. In still other cases, it is due to technological progress and discoveries in an adjoining field, which provide alternative, preferable goods or services, such as natural gas, which has reduced the demand for coal and petroleum, or private cars, which have reduced the demand for buses and trains.

The variety of causes of diminution of demand shows that there is no clear-cut distinction between replacement and cessation of demand as the event which marks the end of an asset's economic life. For example, a generating station has its output reduced as and when other stations can supply electricity more cheaply and it is eventually closed down when there is no longer a demand for electricity at the cost at which it can be operated, but the demand for electricity does not cease. On the other hand, a generating station is not replaced by any particular newer station; rather its potential output is generated by other stations in general, but with no possibility of identifying them because parcels of electricity cannot be labelled. So, although the cost of replacing an asset in some sense is often an important factor in assessing depreciation, it is by no means always the key factor.

DEPRECIATION THEORY

The theory, which comes out of the foregoing considerations and provides the basis for estimating the annual depreciation of non-wasting fixed assets objectively, is that depreciation is the result of just those two causes – physical deterioration of the asset itself and technological progress in the production or provision of the good or service for which the asset is employed, or of a preferred or equivalent substitute for that good or service. Depreciation does not just happen for no particular reason and it cannot be decided by introspection; it is the consequence of recognisable phenomena, which can be observed and assessed.

In general, the total annual cost incurred in respect of an asset has three components:

1) the interest (or loss of other profit) on the net value of the asset;
2) the annual depreciation;
3) the operating cost.

1) and 2) are obviously connected and it is convenient to refer to their sum as the 'capital charge'

The essence of the theory is that the capital charge should be such as to make the total annual cost in every year no greater than the cost of the most economical alternative method of producing or providing the same good or service, or an acceptable substitute.

'Most economical' in this context does not mean 'ideal'. It means the best available estimate of the most economical alternative method. In particular, at the end of the economic life of the asset, its net value cannot be greater or less than its final disposal value, otherwise it would be better to cease using it at a later or earlier time respectively. Also the initial net value at the date of acquisition cannot be less than its acquisition cost because that would imply that its acquisition was uneconomical in known circumstances. (If the acquisition was the outcome of an exceptional opportunity, its initial value may be greater than its acquisition cost, in which case the difference is a windfall gain and should be treated as such in the accounts, but that does not normally happen.)

The theory implies that, between those two dates, the capital charges on the assets employed in the alternative method are to be assessed in the same way as the capital charge on the asset under consideration. But it will become apparent that that does not lead to the nightmare of an incalculable mathematical regression.

In the artificially simple case of no technological progress or diminution of demand, the asset would have to be replaced by an exactly similar asset and both the economic life and the annual depreciation would be determined by the rate at which the asset deteriorated in service. If the annual utilisation were constant, the total annual cost would be constant and the capital charge (as defined) would decrease as and when the operating cost increased, as shown diagrammatically in Figure 1. In as much as the interest must decrease as the net value decreases, the depreciation must also decrease – to zero at the end of the economic life.

The effect of technological progress is to impose a reduction from year to year on the total annual cost, sufficient to ensure that it keeps pace with the ever diminishing total annual cost of the best alternative method. The capital charge – and hence the depreciation – must then decrease correspondingly more rapidly. Both it and the economic life depend on the rate of technological progress as well as the rate of physical deterioration. Moreover the depreciation must commence at a higher level to enable it to decrease annually without reducing the total depreciation over the service life of the asset to less than

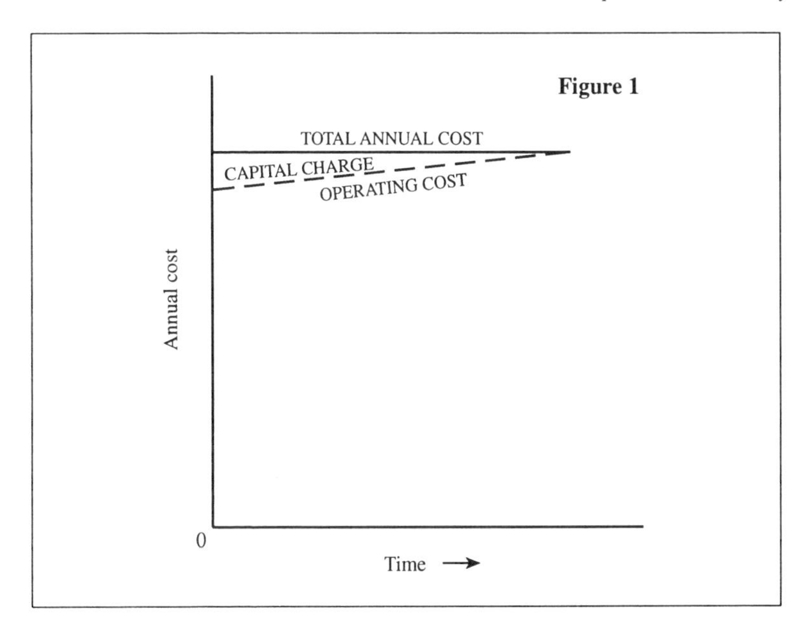
Figure 1
TOTAL ANNUAL COST
CAPITAL CHARGE
OPERATING COST
Annual cost
0
Time

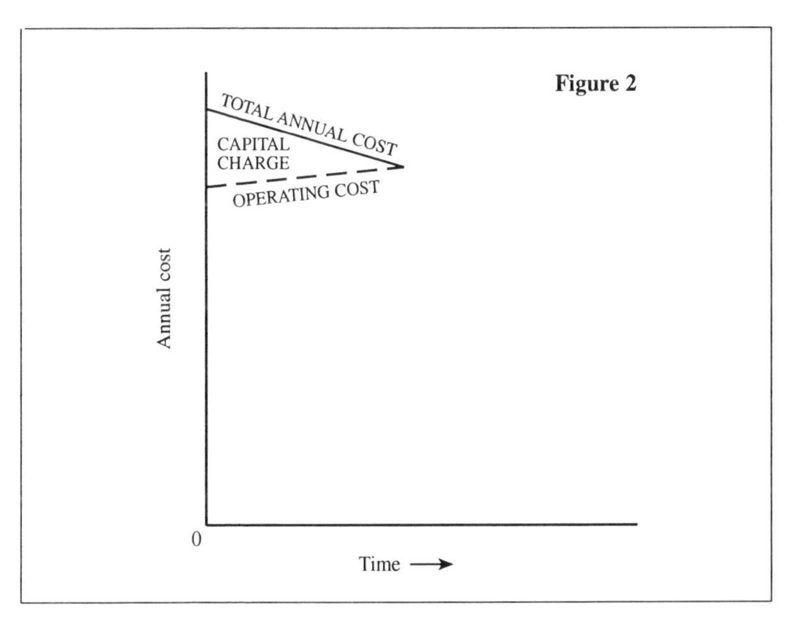
Figure 2
TOTAL ANNUAL COST
CAPITAL
CHARGE
OPERATING COST
Annual cost
0
Time

its acquisition cost (minus its final disposal value, if any), as shown diagrammatically in Figure 2.

At this point, the theory divides into two branches. We have to assume that something in the past will continue in the future, and we can do so without apology because human affairs have always been conducted on that kind of assumption. One would not invest in fixed assets at all without a good deal of foresight and it is only logical to use that same foresight to supplement the principle of Historical Cost where it proves to be inadequate.

The branches are distinguished by the choice of assumptions. The first branch assumes, as does the straight-line method, that fixed assets will have the same economic lives in the future as similar assets had in the past, which leads to a simple approximate method of calculating the annual depreciation during the assumed life, without any additional information except the general knowledge (which the straight-line method ignores) that physical deterioration and technological progress are respectively gradual and gradational. In the second branch, we jettison that assumption and work instead from past rates of physical deterioration and technological progress. On the assumption that those rates will persist, both the economic life and the annual depreciation can be estimated with potentially greater accuracy.

The first branch is short, not too difficult to follow and immediately relevant to annual accounting. The second branch is longer and dependent on expensive information which is unnecessary in most cases but is worth acquiring in some selected cases in order to refine and keep a check on the assumptions on which the first branch relies. Moreover the second branch turns out to be fundamentally important in devising rational, practical systems of pricing, which were stated in Chapter Three (page 51) to be the necessary concomitant of the cost method in designing applications of technology. The exploration of the second branch in this chapter is therefore partly to prepare for Chapters Six and Seven.

The first branch: the wedge formula

The depreciation decreases every year between the pincers of the operating cost and the total annual cost of the best alternative. The economic life ends when they meet, at which point the annual depreciation (not just the net value) must be zero or very small. In the absence of definite information about the best alternative, we cannot say precisely how it decreases; we therefore bridge that gap in our knowledge in the simplest way by assuming that it decreases by the same amount every year. Note, however, that it is the annual depreciation which does so, not the net value.

That is the basis of the wedge formula, so named because the annual depreciation tapers down, like a wedge, to zero at the end of the economic life. To enable it to decrease by the same amount every year while still covering the cost of the asset, it must start at twice the straight-line rate. The arithmetic is then as simple as that of the straight-line method. Figure 3 illustrates the application of the formula to calculate the depreciation of a passenger bus, costing £60,000 with a negligible final disposal value, over a life of 16 years. The depreciation as calculated by the straight-line method is also shown for comparison. The formulae for both methods are given underneath the figure.

With minor variations and successive improvements, the wedge formula has been proposed under various names many times in the last 75 years[5,6,7]. It complies with the requirements of the Accounting Standard, such as they are, but economists and practising accountants have given it the cold shoulder throughout its history[3,8,9,10]. It is an example of the technique of successive approximations, which is widely practised by scientists and technologists but apparently not well understood by accountants.

The distance to the moon provides an example of the technique in astronomy. If one knew nothing of astronomy but had some trigonometry, one might observe the elevation of the moon simultaneously at widely separate points on the earth's surface and, knowing the distances between the points and the

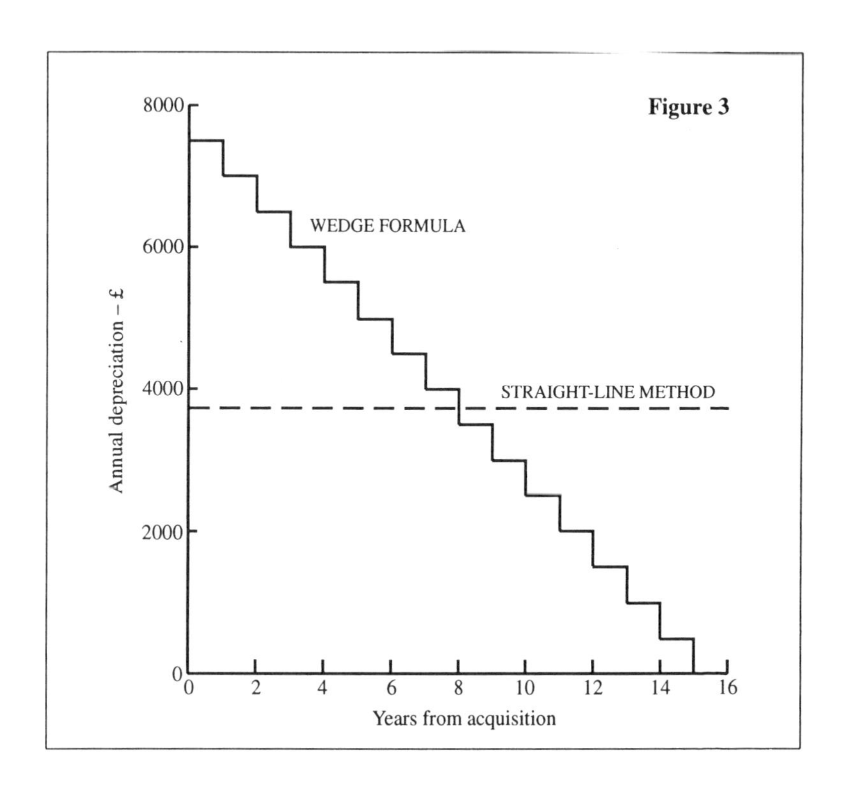

Wedge Formulae

$D = 2(U - F)/Z$
$V = U - D$

Straight-line Formulae

$D = (U - F)/Z$
$V = U - D$

where D = annual depreciation
U = net value at start of year
F = final disposal value (if any)
Z = remainder of economic life
V = net value at end of year

earth's curvature, estimate that the moon circles round the earth at a distance of about half a million kilometres. That would be a first approximation. From their knowledge of Newton's laws of motion and theory of gravitation and some experiments with swinging pendulums, astronomers obtained a much better second approximation. The moon's orbit is not circular but an imperfect ellipse and from those experiments its distance from the earth varies between 356,400 and 406,700 kilometres, with a mean of 384,393 kilometres. The American moon missions provided a third approximation; the reflectors on the moon's surface enable its precise motions, including its excursions from the ellipse, to be calculated with an accuracy of a few metres. Each successive approximation gave a better estimate at the cost of a greater investment of time, expertise and money. The second approximation was not a perfectly precise measurement but it would have been foolish to reject it in favour of the first.

Similarly if one knows nothing of technology but observes that, say, passenger buses commonly have service lives of about 16 years, one might estimate that the depreciation of a bus costing £60,000 is about £3,750 every year. That is a first approximation. A careful study of the causes of depreciation produces a much better second approximation of £7,500 in the first year and decreasing thereafter at the rate of £500 a year to zero in the sixteenth year. That estimate is not as accurate as one would like; it ignores some factors which require much more time and effort to elucidate and are expensive to measure. But that is not a valid reason for rejecting it in favour of the first approximation – as calculated by the straight-line method.

The second branch: predicting the economic life

Predicting the economic lives of non-wasting assets has something in common with predicting the tides. The obvious method of predicting the tides is to observe their ebb and flow and search for a pattern which does not change – and then assume it will continue not to change in the future. That method was probably good enough for primitive fishing communities but its best predictions are unreliable more than a few weeks ahead because the patterns are very indistinct. Better predictions can be obtained from the theory that the tides are caused by two influences, the sun and the moon, whose movements (relative to the earth) conform to a much more definite and consistent pattern. Further improvements can be obtained by observing that the tides are also affected locally by some secondary factors, notably the shape of the coastline and the speed and direction of the wind, which can be measured or predicted separately and brought into the calculations.

Analogously the present practice of predicting the economic lives of assets from the observed service lives of similar past assets is probably good enough for most applications of the cost method because the discounted present values of the designers' options are not very sensitive to their economic lives, as explained in Chapter Three (page 77). But the annual depreciation of an asset is then inversely proportional to its economic life, irrespective of whether the calculation relies on the wedge formula or the straight-line method. In either case an error in the economic life produces an equally large error in the annual depreciation. Better estimates can be obtained by studying the actual causes in more detail. The two causes of depreciation, corresponding to the sun and the moon, are physical deterioration (the minor cause) and technological progress (the major cause).

Thus in estimating the economic life of a new project, we no longer assume that it will be the same as that of previous projects of the same kind. Instead we assume that its rate of physical deterioration and the rate of technological progress in that industry will be the same as they were for previous projects and calculate its economic life from those assumptions. For that calculation, we need to understand what are the secondary factors and how they affect the calculation.

Figure 4 shows the operating cost, interest and annual depreciation of a chemical plant costing £2 million (with a negligible final disposal value), when it deteriorates physically at the rate

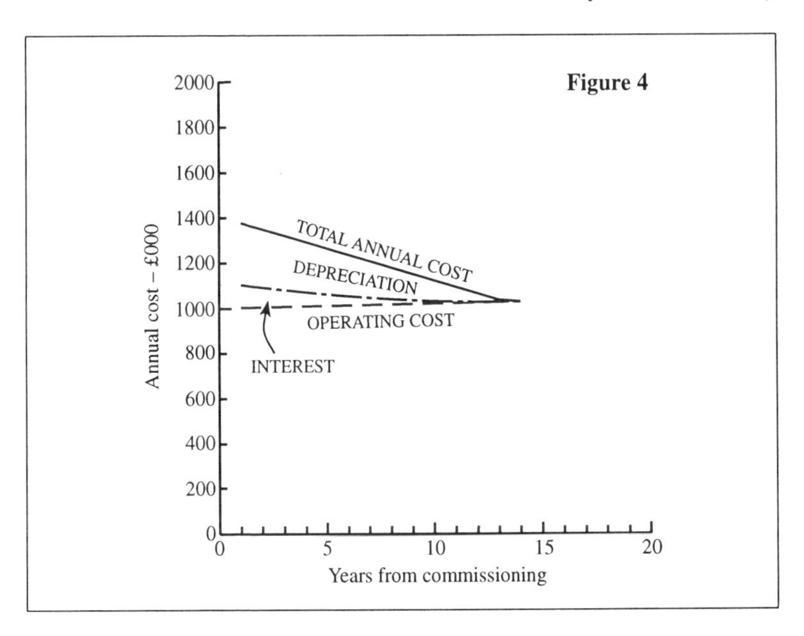
Figure 4
2000
1800
1600
1400
1200
1000
800
600
400
200
0
0
5
10
15
20
Annual cost – £000
Years from commissioning
TOTAL ANNUAL COST
DEPRECIATION
OPERATING COST
INTEREST

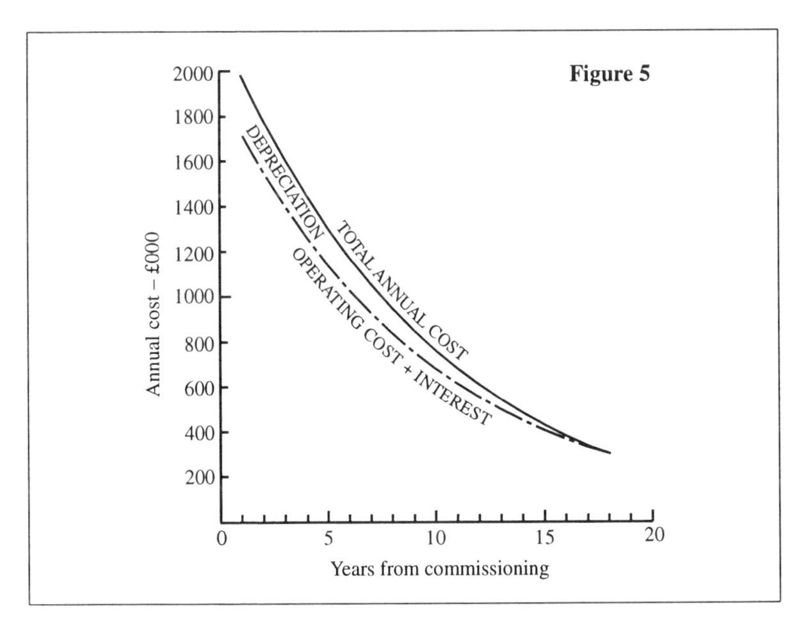
Figure 5
2000
1800
1600
1400
1200
1000
800
600
400
200
0
5
10
15
20
Annual cost – £000
Years from commissioning
DEPRECIATION
TOTAL ANNUAL COST
OPERATING COST + INTEREST

of 0.2% a year and the rate of technological progress is 3% a year. When the output is the same every year, the economic life works out at 14 years. The calculations are explained in Appendix 2.

The secondary factors are the capital cost, the rate of interest and the pattern of utilisation. The effects of the first two can be perceived qualitatively by looking at the figure. The area labelled 'depreciation' represents the capital cost; so, if the capital cost increases, that area will increase, thus increasing the total annual cost of course but also pushing the point at which it comes down to meet the operating cost to the right, that is to lengthen the economic life. Similarly the area labelled 'interest' will increase if the rate of interest is increased, with the same kind of effect on the economic life.

Quantitatively, if the capital cost were £3 million instead of £2 million, the economic life would rise from 14 to 17 years. The same increase would result from increasing the rate of interest from 5 to 15% a year. All three calculations support the view that the wedge formula provides a good approximation to the annual depreciation, once the economic life has been calculated (see Appendix 2), but the more elaborate method is necessary to calculate the economic life in the first place. The straight-line method is clearly inferior to the wedge formula in all three cases.

Figure 5 shows the effect of operating the plant intensively when it is new, to produce 60% more output than in Figure 4, and then reducing the output steadily from year to year. With the same aggregate output over the whole period, that method of operation, which is not at all uncommon, extends the economic life from 14 to 18 years.

Figure 5 also shows, perhaps unexpectedly, that a longer economic life does not necessarily imply a correspondingly lower rate of depreciation in the early years. The rigid connection between those two quantities, which the wedge formula imposes (as does the straight-line method), is broken when the assumption of unvarying economic life is jettisoned. The rate of depreciation and the economic life are still connected in the second branch of the theory, but the connection is weak, particularly when the utilisation varies from year to year. In Figure 5, the depreciation starts at 5% more than the initial rate in Figure 4, but then it declines nearly twice as quickly. The straight-line method is clearly even more irrelevant in such a case. Further details are given in Appendix 2.

It may appear that such calculations merely replace one unknown quantity – the economic life – with two others – the rates of physical deterioration and technological progress. But once the dependence of the economic life on the secondary factors is understood, it should be apparent that direct comparisons of the economic lives of projects operating in different circumstances are apt to be misleading. If the habit can be formed of recording and comparing those rates, greater accuracy and better investment decisions are ultimately achievable – at some additional cost as already mentioned.

In a few cases, where a large organisation owns assets with ages ranging from new to time-expired, the calculations can be more objective, as is explained in Appendix 2. However, the main interest of such calculations lies in their relevance to devising rational systems of pricing – as will be explained in the next chapter.

Some sceptical readers may be thinking that the wedge formula is being justified by hypothetical examples and wondering whether some different examples could be devised to justify the straight-line method. The best way of answering them is to try and work out what kind of hypothetical example, if any, would justify the straight-line method.

The straight-line method implies that the total annual cost suddenly falls at the end of the asset's economic life, but the weight of the evidence is that, except for the tiny minority of assets mentioned in the previous section (page 114), it simply does not. As soon as one realises or admits that physical deterioration is essentially gradual and technological progress is essentially gradational, it follows that, for *any* annual rates of such deterioration and progress, the wedge formula is bound to be a better approximation than the straight-line method, as long as the annual utilisation of the asset is constant or falls from year to year, as it normally does.

The second branch of the theory shows that, if the economic life has been correctly estimated, the greatest errors in the wedge formula occur when the rate of interest is high and the annual utilisation increases from year to year. For example, suppose the annual output – therefore the operating cost – of the chemical process plant in Figure 4, started at half the level in that figure and then increased at the rate of 25% a year for five years, remaining constant thereafter, with interest at the rate of 20% throughout. Then the annual depreciation would be approximately constant during those first five years, giving some apparent credence to the straight-line method, but thereafter it would taper down to zero over the remaining twelve years, as shown in Figure 6. In the absence of such detailed calculations, the wedge formula over the full seventeen years of the economic life of the plant would still represent a better approximation than the straight-line method.

On a point of detail, if an asset is constructed and paid for over a period, its acquisition cost (including interest on payments already made) simply builds up until some utilisation

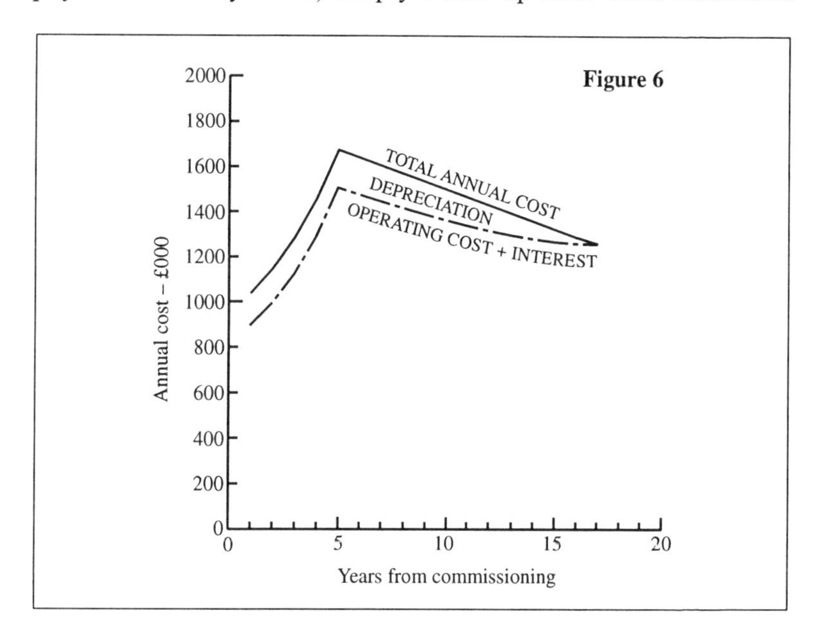

begins. Thereafter depreciation sets in at an annual rate determined by that utilisation and the intended utilisation during the remainder of the economic life, but the net value will go on increasing (or fall more slowly) if there is further capital expenditure on the project (as sometimes happens) after the utilisation has begun. By the same token, if it takes time and money to dispose of the asset when its output or utilisation finally ceases, that cost will reduce the final disposal value; so if it can be estimated in advance it will affect the depreciation during the service life of the asset. Thus, in its most refined form, the theory can include all the factors which could possibly affect the depreciation of any fixed asset by or in which technology is applied. That is not to say that every refinement would be worth its cost, but some would.

In Chapter Three it was mentioned that accountants might legitimately complain that technologists give too little attention to improving the accuracy of their estimates of the service lives of fixed assets, on which they, the accountants, have to rely. The volume of data required for such calculations is too large and the calculations are too complicated for them to be performed routinely for every investment, but the rough and ready estimates which suffice for the discounting calculations in the cost method could be refined by studies, on the foregoing lines, of classes of assets under varying operating conditions. The studies would not be cheap but a possible source of funds and of some of the relevant expertise emerges in the next chapter. Some interest on the part of accountants might expedite the initiation of such studies, but that is not likely to emerge until they have been weaned from the straight-line method onto the wedge formula or something very like it.

THE FORESEEABLE DEPRIVAL VALUE

Figure 7 shows how, under the wedge formula, the net value of the passenger bus in Figure 3 declines to zero over its 16-year life. The net value under the straight-line method is also shown for comparison. The straight-line method overstates the net

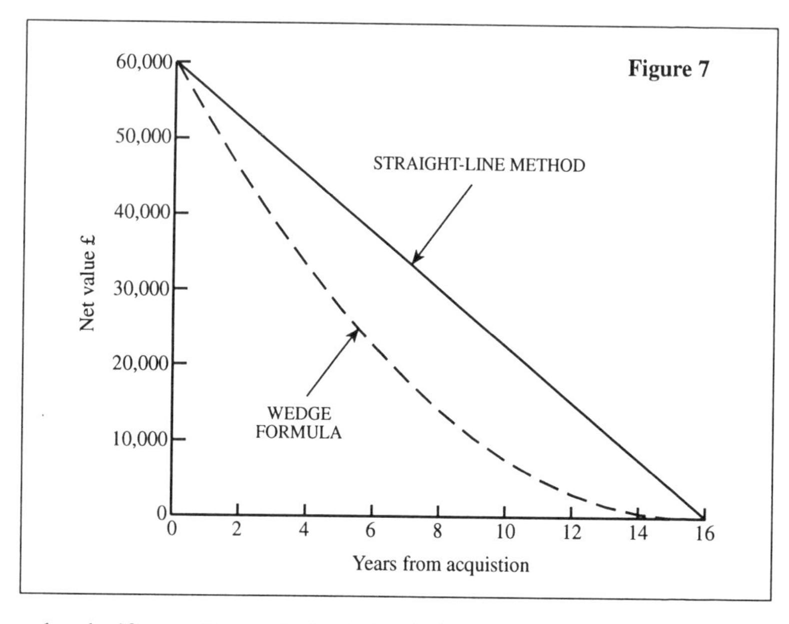

value half way through the life of the asset by more than 100% and the average net value by 50%. Those percentages are not peculiar to buses; they are generally applicable to non-wasting fixed assets by or in which technology is applied. They are too large to be swept under the carpet.

The question is: what is the significance, if any, of those values. To bring it down to earth, let us suppose that a passenger transport company has a fleet of 800 such buses and renews 50 of them every year. Now suppose another such company – perhaps with smaller annual mileages and cheaper fuel and/or maintenance costs – offers to buy 50 eight-year-old buses every year for £1 million. Should our company accept the offer or not?

Under the straight-line method, the total net value of those buses is 50 × £30,000 = £1.5 million. It might therefore appear at first sight that these particular buses are worth half a million pounds more than the offer, which should therefore be refused, but that would not be a valid deduction. It is clear from the Accounting Standard that neither of the conventional methods purports to represent the value of fixed assets to the business in any sense. The £1.5 million is no more than an accounting figure to be carried forward to the start of the next year, with no other significance. And on the rare occasions when such questions arise in real life, the figures in the balance sheets are in fact ignored[11].

On the other hand, under the wedge formula, the total net value of these buses is 50 × £14,000 = £700,000, which is our best estimate of the total amount which can be charged to depreciation over the remaining eight years of the life of those buses without their total annual cost exceeding the total annual cost of the most economic alternative method of providing the same service. It follows that, if those 50 buses were sold for £700,000 and replaced by 50 new buses, costing £3 million, our company's total annual costs would neither rise nor fall. The offer of £1 million therefore represents a potential profit of £300,000 every year.

Of course that is only an estimate and it is only one item of information in deciding whether to accept the offer. The additional administrative costs, the duration of the proposed agreement, the general reputation and credit rating of the buyer and so on would also have to be considered. But all those factors are, or should be, part of any commercial decision. The significance of the £700,000 and the net values of the other 15 batches in the fleet is that each is an estimate of the sum, excluding additional administration costs, which would just compensate our company if it were deprived of the buses in that batch, with full foreknowledge of the deprival. A convenient label for such a sum is the 'foreseeable deprival value'[12].

How could £700,000 be adequate compensation for the loss of fifty buses whose replacements would cost £3 million? The answer is that the difference of £2,300,000 is our best estimate of the discounted present value of the savings we should expect to achieve with the new buses. To simplify the arithmetic we have assumed the new buses will cost the same as the old buses when they were new, but the new buses will have suffered no physical deterioration, so their operating costs will be lower; moreover they will have 16 years of economical operation ahead of them and they will have other advantages, mentioned previously (page 115), stemming from eight years of past technological progress.

We do not know for certain how much lower their operating costs will be, nor the value of their other advantages. Nothing in the commercial world is ever certain. But to the best of our knowledge the old buses, which cost £3 million eight years ago, will be worth nothing in another eight years time, and moreover their rate of depreciation will also be zero by then; *ergo* to the best of our knowledge they are worth £700,000 now, not £1.5 million.

Need it be said that, if we knew more than that – for example if we had recently finished or been given the results of a study which concluded that under our operating conditions the economic life of a bus is not sixteen but twenty years, or that the annual depreciation does not fall at a constant rate – then we should have to amend our calculations accordingly? All estimates depend on the data which go into them and more or better data generally produce better estimates.

It might appear that the reducing-balance method, with a suitable annual percentage reduction, would reproduce the shape of the net value curve under the wedge formula. But what would be a suitable annual percentage?

The normal rate, 25% a year, in the reducing-balance method would reduce the average net value to 53% less than the average under the straight-line method and 30% less than under the wedge formula. And the net value half-way through the life would be reduced by 80% and 57% respectively. However, to bring that about the depreciation in the first year would be £15,000, which implies an implausibly high rate of technological progress, quite at variance with the expected economic life of 16 years. A rate of 16.4% a year would produce the best fit to the wedge formula, but the depreciation would then have to rise from £800 in the fifteenth year to over £4,000 in the sixteenth, which is patently absurd. There can be no point in fitting the reducing-balance method to the wedge formula when the wedge formula is much more realistic and easier to calculate.

So far it has been tacitly assumed that the best estimate of the economic life of an asset at the beginning of its service life will be unchanged throughout that life, but of course any estimate may be overtaken by unforeseen events, which necessitate a reassessment of the remaining economic life or the present net value or both. There is no need to go into the details because they are built into established accounting methods and are not affected if the wedge formula is substituted for the straight-line method. The formula always works from the remaining economic life at the beginning of the year, so changing that estimate involves no more than changing that term in the calculation. If the present net value is affected, the adjustments are more elaborate, but no more complicated than are required for the straight-line method, or any other method.

However, it might be thought that, instead of abandoning the straight-line method, it would be easier just to make the lives of the assets in the calculation less than one's best estimates of their economic lives (or remaining economic lives). For example, one can judge by eye in Figure 6 that, if the life had been taken as ten years instead of sixteen, the straight-line method would have produced much the same net value as the wedge formula (with the full 16-year life) during the first seven years or so, and that thereafter the straight-line value would be consistently less than the wedge-formula value, in fact £7,500 less after ten years. It has indeed been asserted as a matter of belief (though without conclusive evidence) by the compilers of the United Kingdom National Accounts that: 'commercial accountants take a prudent approach in their assessments so that commercial accounts tend to understate the actual lives.'[13]

If that means, as it appears to from the context, that, in applying the straight-line method, commercial accountants habitually understate the expected economic lives, as estimated from the best available evidence, then it implies that they are not complying with their own Accounting Standard. One of the requirements of that standard is that: 'Realistic estimation and regular review of asset lives should result in there being few fully depreciated assets still in economic use. The omission of depreciation on such assets should not be sufficiently material to impair the true and fair view.' A ten-year-old bus with a zero net value in the accounts would be just such an asset still in economic use. There could be no justification for going to such lengths, just to avoid adopting a formula which is as easy to apply as the straight-line method.

The foreseeable deprival value of an asset is not the same as its insurance value, for the reason previously mentioned (page 114) that the purpose of insurance is, or should be, to cover events whose timing and precise nature are unforeseeable, whereas depreciation is essentially foreseeable. However the foreseeable deprival value is a good starting point for estimating the insurance value of a non-wasting asset. If we lose an eight-year-old bus, with a foreseeable deprival value of £14,000, in an accident, we expect to replace it by a new bus, but meanwhile we shall suffer additional expense. If we estimate that that will amount to, say, 20% of the foreseeable deprival value of the old bus plus £2,000, the insurance value is £14,000 + 20% + £2,000 = £18,800. Neither the straight-line method nor the declining-balance method provides any starting point for such an estimate.

SUMMARY

The fixed assets, so-called, by or in which technology is applied form the greater part of a modern society's accumulated wealth. The straight-line method of calculating the annual depreciation of non-wasting fixed assets grossly understates it in the early years of their economic lives and overstates it correspondingly in the later years. Consequently the almost universal use of the method by companies and undertakings (including non-profit-making concerns) decimates the utility of their annual accounts as sources of information for important decisions by managers, shareholders and other interested parties. It distorts the prices charged for the goods and services produced and provided by those concerns. It also inhibits the instigation of co-operative studies of the economic lives of tangible fixed assets. On all those counts it reduces the contribution of technology to prosperity.

The straight-line method has those deficiencies because it ignores the known causes of depreciation of most non-wasting fixed assets, namely their gradual physical deterioration and the essentially gradational progress of technology. The reducing-balance method, which also ignores those causes, has corresponding deficiencies.

The theory of depreciation described in this chapter would eliminate those deficiencies. The wedge formula would remove most of them without any further research. Its arithmetic is as simple as that of the straight-line method.

The straight-line method also grossly overstates the average net values of non-wasting fixed assets during their service lives. The net values which come out of the wedge formula are estimates of the foreseeable deprival values of the assets. That advantage will be seen to be particularly important when inflation rears its ugly head.

The second branch of the theory of depreciation described in this chapter provides a possible means of improving the estimates of the economic lives of assets under different circumstances. It also turns out – perhaps unexpectedly – to be fundamentally important in devising rational, practical systems of pricing, which were stated in Chapter Three (page 51) to be necessary concomitants of the cost method in designing applications of technology. Those matters are the subjects of Chapter Six.

REFERENCES AND FOOTNOTES

1) Mr. Micawber in *David Copperfield* by Charles Dickens. For readers unfamiliar with the British money system before decimalisation, nineteen nineteen six would be £19.975 and twenty pounds ought and six would be £20.025 in the decimal notation
2) Strictly the term 'fixed assets' also includes a company's investments in other companies, including even such items as goodwill, but in as much as such investments normally represent tangible

fixed assets belonging to the other companies, there is no need to deal with them separately.

3) Since the middle of the last century according to F. and V. Lutz in *The Theory of Investment of the Firm*, Princeton University Press, 1951.
4) *The Philosophy of Physical Science* by Sir Arthur Eddington, Cambridge University Press 1939. For readers unfamiliar with Imperial measures, one mile is about 1.6 kilometres and there are – or were before metrication in Britain – 1,760 yards in a mile. The accepted theory comprises Newton's laws of motion and theory of gravitation. The italics in the quotation are Eddington's own.
5) H. Hotelling, A General Mathematical Theory of Depreciation, *Journal of American Statistical Association*, Volume 20, 1925. His formula took account of physical deterioration, operating costs, varying utilisation and the rate of interest (see Appendix 2) but it made no allowance for technological progress.
6) E. J. Liljeblad, *Depreciation of Industrial Plant*, Vaestmanlands Allehanda Printing Co, Sweden, 1939. His formula included the effects of technological progress but not those of varying utilisation.
7) D. Rudd, 'A General Theory of Depreciation of Engineering Plant', *Proceedings of the Institution of Electrical Engineers*, Vol. 108, Part A, no. 41, October 1961.
8) E. L. Grant & P. T. Norton Jr., *Depreciation*, Ronald Press Co., 1949. 'There is no one right method of distributing the difference between the first cost and salvage value among the years of life...'
9) T. J. Morgan, *Telecommunication Economics*, Macdonald, 1958. 'The determination of the value of an existing asset during its life is quite artificial and arbitrary.'
10) J. M. Drummond, Chief Financial Officer of the Central Electricity Generating Board, at a discussion of Reference 7 above at the I.E.E. in 1961. 'In my view there is no such thing as a precise measurement of depreciation over one year; it is an arbitrary assessment.'
11) R. S. Edwards & F. H. S. Brown, 'The Replacement of Obsolescent Plant', *Economica*, August 1961. They were respectively Chairman of the Electricity Council and Deputy Chairman of the Central Electricity Generating Board at that time.
12) See A. J. Merrett & A. Sykes, 'Investment in Replacement: The

Optimal Replacement Method', *Journal of Management Studies*, May 1965. They defined the deprival value as 'the sum of money which would just compensate the firm for its loss in stated conditions, given the action that will be taken by the firm to minimise this loss', but they did not distinguish between accidental and foreseeable deprival.

13) 'United Kingdom National Accounts – Sources and Methods', Central Statistical Office, HMSO 1985

APPENDIX 1 – ACCOUNTING METHODS OF CALCULATING DEPRECIATION

The accountants' definition of the problem was stated as follows (not for the first or last time of course) by F. K. Wright in 'Towards a General Theory of Depreciation', *Journal of Accounting Research*, vol. 2 # 1, Spring 1964, page 81:

> For a specific asset, objective verifiable values based upon external transactions are available at only two points in time: at the moment of acquisition, and at the moment of disposal. If these two events occur within the same accounting period, no problem arises. But where the events are widely separated in time (as is usually the case with fixed assets), determination of periodic income is impossible without establishing a value for the asset at the end of each intervening period. The problem of depreciation accounting is the problem of establishing these needed values without the objectively verifiable basis which only external transactions can provide.

We must first clear up a possible confusion of meanings. In that passage 'periodic income' does not have its ordinary meaning of what comes in – the 20 pounds in Mr Micawber's happiness/misery formula. It means the ordinary income minus the expenditure (including interest) and the depreciation in the period of account – in other words the profit in that period.

The depreciation is not strictly an item of expenditure since it does not involve any external transaction, but it has to be subtracted, along with the expenditure items, from the income in the ordinary Micawber sense in calculating the profit.

Accountants and others sometimes try to avoid the word 'profit' because some people tend to assume that any organisation which makes a profit must be attempting to maximise that profit and they object to that motivation on moral grounds. But the assumption is invalid in that the accounting methods, and in particular the problem of annual depreciation, are not affected by whether the motive is to maximise the profit or eliminate it or anything in between. In this book it is never the intention to provoke controversy for its own sake but, where the choice must be made, the risk of controversy is preferred to the risk of confusion; so 'income' is not used as a euphemism for 'profit'.

In the eyes of a technologist, the foregoing definition is an over-simplification in as much as large fixed assets, such as generating stations, do not have definable moments of acquisition, nor are their acquisition values always based solely on external transactions. They are constructed and paid for gradually over periods of several years, and part of the cost is often that of the owner's own design staffs, who engage in more than one project at a time and whose remuneration, accommodation and other employment costs must therefore be allocated by internal analysis. Moreover design and construction faults sometimes reduce the capacities of stations for years into their service lives, and so make even the moment of commencement of commercial operation (which would be an acceptable substitute for the moment of acquisition) incapable of unequivocal determination.

However these are minor aspects. Our understanding of the central problem will not be seriously weakened if we begin by thinking of every fixed asset as having an objectively verifiable acquisition cost at an objectively verifiable moment. The adjustments for assets which are constructed and come into operation gradually can be added later. Furthermore, we can simplify the problem without serious error in most cases by ignoring the final disposal value or relegating it to a parenthesis because it is normally very small.

In Britain the knowledge and wisdom of accountants are summarised in a set of 24 Statements of Standard Accounting Practice (SSAPs) (no. 24 was issued in 1988). No. 12 is 'Accounting for depreciation'. It was issued by the Accounting Standards Committee in 1987 and revised in 1988 and it supersedes the old Recommendation N9, 'Depreciation of fixed assets', which came out in 1945.

SSAP12 defines depreciation as 'the measure of the wearing out, consumption or other reduction in the useful economic life of a fixed asset whether arising from use, effluxion of time or obsolescence through technological or market changes'. The odd wording might lead you to suppose that the depreciation and the life of an asset are reckoned in the same units, but that would be wrong; the life is reckoned in years and the depreciation (in Britain) in pounds sterling. SSAP12 requires no more than that the accounts shall disclose 'the depreciation method' (that is the method of calculating the annual depreciation) and the effect of any change in the method and the reason for that change, as mentioned in the main text.

Recommendation N9 was less permissive. It stated that there were several methods but that 'Provision for depreciation [of non-wasting fixed assets] should, in general, be computed on the straight-line method', also as described in the main text. It is so called because, if you plot a graph of the net value of the asset at the end of each successive year, the points all lie along a straight line, which falls from the acquisition cost at the moment of acquisition to zero (or the disposal value if it is not negligible) at the end of the expected service life (referred to in N9 as 'the period of anticipated use').

N9 also described the 'reducing-balance' method, but did not recommend it. The tax authorities in Britain use that method at the time of writing (1998) for calculating liability for tax, with a fixed percentage of 25% a year, for plant and machinery. It is administratively convenient because the calculation does not depend on the economic life, so it can be applied indiscriminately at the same rate to many kinds of fixed asset without regard to their differing economic lives. 'Plant and machinery' is deemed to include motor vehicles and other things which are not plant or machinery in ordinary parlance. Mathematically minded readers will realise that the net value then declines exponentially from year to year; it never falls to zero and it is generally greater than the final disposal value (if any) at the end of the economic life – much greater for assets with short lives. A correction must therefore be made in the final year, which increases the cost arbitrarily at that time. Incidentally, the tax authorities prefer the term 'writing-down allowance' to 'depreciation', but it is a distinction without any real difference.

Small and large companies

In a small company with only a few fixed assets, the acquisition and possible replacement of which are major events, the chosen method determines how soon or late the profit or loss from the ownership of those assets will appear in the annual accounts, so obviously the choice of method is important in a small company. But what about a large company or undertaking with a more or less continuous programme of investment in new fixed assets? Does it matter much which method such a company uses?

For an example, let us consider a public transport company which owns a fleet of 800 buses, all having the same passenger capacity and performance (speed, acceleration, braking power, etc), costing £60,000 each and all having a service life of 16 years. As long as the fleet stays the same size and the buses have the same price and service life, the company can arrange its investment programme to take 50 buses out of service every year and replace them with 50 new buses at a cost of £60,000 × 50 = £3 million. Using the straight-line method, the annual depreciation on every bus will be £60,000/16 = £3,750, so the total annual depreciation will be £3,750 × 800 = £3 million, that is the same as the cost of the new buses. Which other methods might be theoretically contemplated?

At one extreme the whole cost of every new bus might be counted as depreciation in the year in which it is acquired, producing a total of £60,000 × 50 = £3 million again. At the other extreme, that attribution might be postponed until the last year of the service life, but that would also produce the same answer. The reducing-balance method would also lead to a total annual depreciation of £3 million, whatever the fixed annual percentage; in fact no method with any pretence of rationality, if applied consistently, could produce a different result. In such a large company it seems not to matter much, in calculating the annual profit or loss, which method the accountants choose, right out to those extreme limits in either direction.

However large companies are not eternal; they have to grow from small beginnings and eventually they have to shrink again, one way or another. Suppose the fleet has been expanding steadily at the rate of 10% a year. Then 16 years ago it was only 174 buses strong and only 20 buses were acquired at the start of that year of account. Since then the number in each batch of new buses has increased every year until the last batch had to be 93 to make up the total of 800 buses for the year just ended (73 for the 10% increase from 727 at the end of last year, plus 20 to replace the batch acquired 17 years ago).

Now by the straight-line method the depreciation in the year just ended is still £3,750 per bus throughout the fleet, i.e. £3 million. But by the reducing-balance method, at the fixed rate of 25% a year, it starts at 25% of £60,000 = £15,000 per bus for the newest, largest batch of 93 buses, then falls to 25% of £(60,000 – 15,000) = £11,250 per bus for the second newest batch (85 buses), and so on down descending scales of both depreciation per bus and buses per batch. That bias towards higher depreciation in the newer, larger batches produces a total for the fleet of £4.4 million – 47% greater than by the straight-line method. And a growth rate of 10% a year is not out of the way; if it were 20%, the total for the 800-strong fleet by the declining-balance method would be over £5.6 million, that is nearly 90% greater than by the straight-line method. (The extreme methods mentioned in the earlier paragraph would lead to even larger differences, compared with the straight-line method, but common sense would preclude them; their only purpose was to show how insensitive the profit-and-loss accounts are to the choice of method when the size of the fleet is not changing.)

If the acquisitions for such an expansion are financed by issuing ordinary shares, the shareholders will evidently receive their profits earlier under the straight-line method than under the reducing-balance method, but if the company makes losses the shareholders will not feel them until later. In other words the shareholders in a large expanding company are affected by the method of calculating the annual depreciation in much the same way as those of a small company which has recently increased its investment in fixed assets. More importantly perhaps, the prices to yield a particular profit – or none – will be substantially lower under the straight-line method than under the reducing-balance method. But when such a company ceases to expand and is obliged to contract, which may happen unexpectedly and sometimes suddenly, then the prices to yield a profit (or avoid a loss) will be correspondingly higher under the straight-line method, just when the company's financial position is likely to be particularly weak.

Those considerations in themselves do not prove that the reducing-balance method is better than the straight-line method, nor indeed that any particular method is better or worse in any sense than any other, but they do show that the choice of method has practical consequences under circumstances which are bound to occur, even in a large company which acquires new assets every year. Those consequences bring us back to the questions: can annual depreciation be measured or estimated and, if so, how? If it cannot, the accounts can still function to prevent embezzlement, comply with legislation and so on, but their utility as a source of information for decisions by managers, shareholders and other interested parties is decimated. A set of accounts for our public transport company which shows a profit of, say, £1 million does not provide an objective basis for important decisions if it could equally well have shown a loss of £1.4 million, merely by choosing a different method of calculating one of its main components – and with no reason disclosed for having chosen one method rather than another in the first place.

The choice of method has equally important consequences when the acquisition costs, service lives and/or operating costs of the assets change, irrespective of whether the business is expanding, contracting or static. Indeed one of the principal aims of technology is to make things more cheaply and sometimes to make them last longer as well.

Wasting assets

The important difference between wasting assets (mines, oil wells and quarries, etc.) and non-wasting assets is that the former are, in a real sense, consumed; moreover their annual consumption can be measured. SSAP12 gives no guidance on how their annual depreciation should be calculated, but N9 said: 'Provision for depreciation and depletion should be made according to the estimated exhaustion of the asset concerned', and that is the normal practice. A technologist can have no quarrel with that kind of recommendation.

That recommendation did not specify explicitly how the value of a wasting asset should be decided at the commencement of extraction. It should obviously include the costs of acquiring the site and preparing it for extraction, but there is also the scarcity value of the mineral to be considered. However the problems of scarcity valuation are quite different from those of estimating annual depreciation, once the initial scarcity value has been determined. Scarcity valuation was touched on in Chapter Three (page 93), but it will not be tackled directly until Chapter Eight.

APPENDIX 2 – THE SECOND BRANCH OF DEPRECIATION THEORY

How can physical deterioration be measured or estimated? The rate at which the operating cost increases has some relevance if it increases steadily, but it may not do so. If the annual mileage of a bus decreases from year to year, its operating cost will also decrease although it is deteriorating, and the same applies to generating stations and fixed assets in general, but the operating cost per kilometre (or per kilowatt-hour) will normally increase and that increase is a potential measure of the deterioration. However, as many motorists know, the fuel consumption (which accounts for a large part of the operating cost) of any motor vehicle depends also on its speed and the traffic conditions, which may vary from journey to journey and must therefore be sieved out to measure its deterioration. Similarly for other assets, the operating cost depends on the conditions in which it is used.

Estimating the rate of technological progress comes up against the same problems because the total annual cost of the best alternative method of producing or providing the good or service depends on the same conditions. It would be misleading to pretend that either estimate is not inherently complicated, but it would also be misleading to leave the impression that they cannot be estimated at all. There are probably no more than four or five physical measures of the annual utilisation of a generator in a thermal station (as defined in Chapter 2, page 20) which are important enough to be regarded, namely:

- i its output in kilowatt-hours;
- ii (sometimes) its output in excess of its most efficient output (that is the output at which its fuel consumption per kilowatt-hour is a minimum);
- iii the number of hours it is connected to the grid, irrespective of its output;
- iv the number of times it and its boiler and turbine are started up and shut down, irrespective of its output or the durations of the periods of connection and disconnection;
- v the number of hours during which it is available to be started up on demand, irrespective of the first three measures.

The whole operating cost can be resolved with sufficient accuracy into six components, five of which are proportional to those five measures and the sixth is independent of the utilisation. The resolved costs per kilowatt-hour, per kilowatt-hour in excess of the most efficient output, per hour of operation, per start-up/shut-down and per hour of availability are known as 'cost parameters', and the rates at which they increase from year to year are measures of physical deterioration. Similarly the total annual cost of the best alternative can be resolved into corresponding components with corresponding parameters, and the rates at which all six of those parameters decrease from year to year are measures of technological progress. For other types of asset, the measures and the corresponding cost parameters are of course different, but the number of measures which are important enough to be taken into account is unlikely to exceed five or six and often a single measure or a pair will be sufficient.

The relevant quantities are the average rates at which the parameters of the operating cost increase and the parameters of the total annual cost decrease, ignoring the accidental, unpredictable fluctuations in those rates in individual years. For ease of presentation, let us begin with single assets assessed on their own, and then pass on to the special opportunities offered by groups of assets which are employed to produce or provide a common good or service.

The secondary factors

Referring to the chemical plant in Figures 4, 5 and 6, if X, Y and Z denote its total annual cost, its operating cost and its capital charge respectively, then of course $X = Y + Z$. If the utilisation is constant, X will increase slightly from year to year, due to physical deterioration. If f is the annual rate of that increase and the suffix m denotes the mth year of operation, then:

$$Y_{m+1} = Y_m.(1 + f) \qquad so \qquad Y_{m+1} - Y_m = f.Y_m$$

Similarly, if g is the annual rate of technological progress, then:

$$X_{m+1} = X_m.(1 - g) \qquad so \qquad X_m - X_{m+1} = g.X_m$$

Whence

$$Z_m - Z_{m+1} = X_m - Y_m - X_{m+1} + Y_{m+1} = g.X_m + f.Y_m$$

Hence the whole series of capital charges can be calculated iteratively by varying the capital charge in the first year, Z_1, until the net value falls from the initial cost to the final disposal value by the time the annual depreciation falls to zero. That is how Figure 4 was constructed and how the economic life was recalculated for a capital cost of £3 million instead of £2 million and for a rate of interest of 15% instead of 5%.

If the utilisation varies from year to year, the operating cost and the total annual cost will vary to reflect those variations as well as the effects of physical deterioration and technological progress. Suppose there is only one measure of utilisation (e.g. tonnes of annual output), denoted by Q, and one cost parameter, denoted by P, so that $Y_m = P_m.Q_m$.

Then

$$P_{m+1} - P_m = f.P_m$$

And

$$\begin{aligned} Y_{m+1} - Y_m &= P_{m+1}.Q_{m+1} - P_m.Q_m \\ &= P_{m+1}.Q_m - P_m.Q_m + P_{m+1}.Q_{m+1} - P_{m+1}.Q_m \\ &= f.Y_m + P_{m+1}.(Q_{m+1} - Q_m) \end{aligned}$$

The first term, $f.Y_m$, on the right-hand side of that equation represents the effect of the physical deterioration and the second term represents the effect of the change in utilisation. The latter effect must of course be reflected in the total annual cost, so that:

$$X_m - X_{m+1} = g.X_m - P_{m+1}.(Q_{m+1} - Q_m)$$

And

$$\begin{aligned} Z_m - Z_{m+1} &= X_m - Y_m - (X_{m+1} - Y_{m+1}) \\ &= X_m - X_{m+1} + Y_{m+1} - Y_m \\ &= g.X_m + f.Y_m \qquad \text{as before.} \end{aligned}$$

In other words, it is not necessary to know the utilisation and the cost parameter separately to calculate the capital charge. The same is true if there are several parameters, the only proviso being that they must all be equally affected by physical deterioration and all equally affected by technological progress. Thus the same formula can be used whether the utilisation varies from year to year or is constant – and it was used to construct Figures 5 and 6.

If it is known that a particular item of cost – perhaps the cost of some personnel or part of the overhead charges – is not affected or is hardly affected by those causes of depreciation, then that item should be omitted from the calculation. In practice some judgement may be necessary to decide which items to include and which to exclude. As always, the accuracy of the calculation depends on the quantity and quality of the available data.

It may perhaps be surmised that in some circumstances there may be several cost parameters which are affected differently by technological progress. If there are any reliable data behind that kind of surmise, they should of course be brought into the calculation, but otherwise it will be more accurate to assume that the parameters are all equally affected than to retreat into an arbitrary method of assessing the capital charge.

Groups of assets

A more elaborate but less subjective method of estimating the economic lives of assets is possible in some large companies or undertakings which have more or less continuous programmes of investment in groups of assets. If the assets in such a group are employed in the production or provision of a common good or service and they have a range of ages from new to obsolete, then the data for the calculations can be obtained from the internal accounts – provided of course that the accounts are kept with that purpose in mind.

The obvious example is a group of generating stations in an electricity supply system. For the greatest accuracy, the group should be as large as possible. The generating stations in England and Wales before the privatisation of the supply industry in 1990 were almost ideal in that respect because they were all owned by one undertaking, the Central Electricity Generating Board. (However that is not to say that single ownership is or is not ideal in other respects.)

The first step is to express the operating cost, Y, of any asset in the group as:

$$Y = P_1.Q_1 + P_2.Q_2 + P_3.Q_3 + P_4.Q_4 + P_5.Q_5 + S$$

where P_1, P_2, ... P_5 are the five cost parameters,
Q_1, Q_2, ... Q_5 are the corresponding measures of its output or utilisation,
S is the sixth component, which is independent of the output or utilisation.
and the total annual cost, X, is given by:

$$X = Y + Z$$

where Z is the capital charge, as in the case of a single asset which is not in such a group.

The method is to compare the operating cost of any asset with the cost of obtaining its output from other assets in the group. To enable that comparison, we must first allow for the different capacities of the individual assets (unless they all happen to have the same capacity). The capacity of a generator is the maximum output, C (measured in kilowatts), which it can sustain for prolonged periods – several hours; the capacity of a bus is the number of passengers it can carry, and so on. The revised equations are then:

$$y = P_1.q_1 + \ldots + P_5.q_5 + s \quad \text{and} \quad x = y + z$$

where

$$y = Y/C,\ q_1 = Q_1/C \text{ etc., } s = S/C,\ x = X/C \text{ and } z = Z/C.$$

Then if the suffixes a and b denote any two assets in the group, the theory requires that ideally x_a should not exceed the cost of obtaining the output q_a from the asset b and vice versa, i.e.

$$x_a = P_{1a}.q_{1a} + \ldots + P_{5a}.q_{5a} + s_a + z_a <= P_{1b}.q_{1a} + \ldots + P_{5b}.q_{5a} + s_b + z_b$$

and

$$x_b = P_{1b}.q_{1b} + \ldots + P_{5b}.q_{5b} + s_b + z_b <= P_{1a}.q_{1b} + \ldots + P_{5a}.q_{5b} + s_a + z_a$$

which can be rewritten as:

$$\Delta P_1.(\mu q_1 + \tfrac{1}{2}.\delta q_1) + \ldots + \Delta P_5.(\mu q_5 + \tfrac{1}{2}.\delta q_5) => \Delta s + \Delta z => \Delta P_1.(\mu q_1 - \tfrac{1}{2}.\delta q_1) + \ldots + \Delta P_5.(\mu q_5 - \tfrac{1}{2}.\delta q_5)$$

where

$$\Delta P_1 = P_{1b} - P_{1a}, \text{ etc.} \qquad \mu q_1 = \tfrac{1}{2}.(q_{1a} + q_{1b}), \text{ etc.}$$
$$\delta q_1 = q_{1a} - q_{1b}, \text{ etc.} \qquad \Delta s = s_a - s_b \text{ and } \Delta z = z_a - z_b$$

Whence

$$\Delta z = \Delta P_1.(\mu q_1 \pm \tfrac{1}{2}.\delta q_1) + \ldots + \Delta P_5.(\mu q_5 \pm \tfrac{1}{2}.\delta q_5) - \Delta s$$

In that equation, $\tfrac{1}{2}.(\Delta P_1.\delta q_1 \pm \ldots \Delta P_5.\delta q_5)$ represents a residual uncertainty. It cannot be eliminated but it can be minimised by listing the assets in order of decreasing utilisation and then comparing every adjacent pair in the list. If one measure of the utilisation is dominant (e.g. for electricity generators the output, q_1), that listing is straightforward, but in some cases it may be worthwhile to re-arrange some of the pairs to reduce the uncertainty a little more. Then the equations for the mid-position are:

$$\Delta z = \Delta P_1.\mu q_1 + \ldots + \Delta P_5.\mu q_5 - \Delta s$$

Hence

$$\Delta x = \Delta y + \Delta z = \mu P_1.\delta q_1 + \ldots + \mu P_5.\delta q_5$$

The former equation enables the ideal capital charge on – and hence the depreciation of – every asset in the group to be calculated from the charge on any one of them, and so (by iteration) from the total capital charge, but the equations do not yield the total capital charge, which must therefore come from a different source. The obvious source is the sum of the charges on the individual assets, as calculated from their 'standard' economic lives using the wedge formula, but bearing in mind that the charges on the individual assets from the above equations will not then be the same, asset by asset, as the charges by the wedge formula. To save words, let us call the former charges 'comparative charges' and the latter 'accounted charges'.

Those discrepancies arise because the wedge formula does not take account of the effects of the capital cost, rate of interest and utilisation pattern of an asset on its economic life. The discrepancies therefore indicate the extent of those effects. The comparative capital charge represents the amount which could be charged to an asset without its total annual cost exceeding the cost of the most economic alternative method of producing or providing that particular good or service. If it is greater than the accounted charge, that asset is more economic – at its particular output or utilisation – than the adjacent assets in the list. Its net value was therefore higher and its remaining economic life was longer at the beginning of that year than were shown in the accounts of the previous year – and vice versa if the comparative charge is less than the accounted charge.

The latter equation (for Δx) enables what we may call the 'economic cost' of every asset in the group to be calculated. It is the sum of the operating cost and the comparative capital charge and it represents the total annual cost of the most economic means of producing or providing that good or service which would provide *in toto* for the depreciation of all the assets in the group in that year.

It might appear at first sight that the accounted capital charge on the former assets (whose comparative charges are greater than the accounted charges) should be increased to reduce the discrepancy, but that would be counter-productive because it would reduce the net value at the beginning of the next year and so increase the discrepancy in that year. And it would not normally be appropriate to increase the net value of those assets at the beginning of the current year because that would invalidate the accounts of the previous year unnecessarily. Rather the response should be to lengthen the remaining economic lives of the former assets at the beginning of the current year and, if necessary, reduce the capital charge on them in the current accounts, so as to increase their net value at the end of the year – and vice versa for the latter assets.

By such adjustments, the two sets of charges could be gradually brought more nearly into line with each other, but it is to be expected that the discrepancies would fluctuate from year to year. So it is not suggested that the discrepancies could be eliminated in a single year, nor that the wedge formula should be abandoned. The main text has shown that the wedge formula is a good approximation, whatever the pattern of utilisation, if the economic life has been well estimated. The value of the comparative equations lies in the information they provide about the relative economic lives of the assets in the group, without any subjective judgement of the rates of physical deterioration or technological progress. Those rates are implicit in the data because the younger assets will have deteriorated less and benefited more from technological progress than the older assets in the group, but they are not explicit and need not be made so.

But what about the absolute economic lives of the assets and the total capital charge? Is there not an arbitrary element in the calculation of the total accounted charge from the standard economic lives of the assets? Yes there is, and it could be gradually reduced by further attention to the comparative equations.

As the discrepancies between the comparative charges and the accounted charges are reduced, it will be gradually revealed that the majority of the assets are reaching the ends of their economic lives before – or after – their capital costs have been fully redeemed by the respective accounted charges. In the former case, the total capital charge is too small and should be increased, and of course in the latter case it is too large and should be reduced.

Thus over a number of years the comparative equations could be used to recalculate both the economic lives of the individual assets, to take account of the primary and secondary factors affecting their depreciation, and the total capital charge, to take account of the physical deterioration and technological progress of the whole group. Any further analysis to estimate the actual rates of physical deterioration and technological progress would be for the purpose of estimating the economic lives of separate single assets which are not in the group. It would not serve any very useful purpose so far as the assets within the group are concerned.

A numerical example of the application of the comparative equations in one year was provided in the author's paper on 'A General Theory of Depreciation of Engineering Plant', which was cited in the main text of this chapter (Footnote 7). A more comprehensive example, covering a longer period, would require a large company or undertaking with a suitable investment programme and range of assets to keep its accounts with that purpose in mind.

5

ANNUAL ACCOUNTING FOR MONETARY INFLATION

There's none so blind as they that won't see.[1]

Once upon a time long ago, before Caesar came to Britain, merchants from Phoenicia on the Eastern Mediterranean seaboard used to sail to Cornwall to sell carpets and buy tin, which was mined in that region. The merchants lost some fine ships and crews and valuable cargoes off the Cornish coast in mysterious circumstances. When their sailors thought they were a safe distance from the coast and sailing away from it at dusk, they were nevertheless sometimes wrecked on the rocks before dawn.

The sailors navigated with the aid of the log-and-line method to measure the speed of their ships. It is simple. You tie knots at regular distances along a long line of rope; then you tie a log of wood to the end of the line and throw the log into the sea over the stern of the ship. If you keep the line slack, the log will lie still in the water and you can count and time the knots as they pass over the stern. That gives you the speed of your ship with reasonable accuracy when you are sailing in the Mediterranean Sea, but it can be and was terribly misleading in the English Channel.

Some lucky sailors, who escaped from being wrecked and returned to Phoenicia, spoke of an evil influence, never encountered in the Mediterranean, which they had heard the Britons refer to as 'Tide'[2]. They recounted with horror how a ship could be making good progress away from the rocks, according to their measurements, and yet still be carried onto those rocks by the dreaded Tide. A few of the more thoughtful and articulate sailors spoke of a difference between the speed of a ship through the sea and its true speed, which difference, they said, represented the speed of the Tide. They claimed that they could actually measure the speed of the Tide and that, by setting their courses by sight, wind and Tide instead of just by sight and wind, they could reach their destinations more directly and less dangerously.

After studying the sailors' log books and cogitating for four years, The Council of the Institute of Chartered Navigators in Phoenicia and Lebanon denied that the Tide could be 'quantified', which was a fashionable way of saying that its speed could not be measured. They insisted that the ships had been driven off their courses by storms and they proposed a radically different and complicated navigation system, which they claimed would work better than the traditional system in storms, quite ignoring their sailors' proven ability to navigate successfully, with nothing more elaborate than a log of wood and a knotted line, through the storms which raged, then as now, from time to time in their familiar (non-tidal) home waters.

After much argument and several modifications, which added to the complexity of the new system, it was imposed on the sailors for nine long years, until its futility became so painfully apparent that it was suddenly withdrawn. But the perpetrators offered nothing in its place even then. They persisted that their navigation handbook was still an authoritative reference work. The sailors were abandoned to their fate and the wrecking propensities of the tides. No-one knows for sure how the story ended but the sailors did not live happily ever after.

Substitute 'accounting' for 'navigation', 'monetary inflation' for 'Tide', 'the Historical Cost principle' for 'the log-and-line method' and 'England and Wales' for 'Phoenecia and the Lebanon' and you have something like a potted history of the attempts of the accounting profession in Britain to account for monetary inflation up to 1988. But it is not a fairy tale (unlike the story of the Phoenician traders), nor are its main features peculiar to Britain, although the details differ in other countries. No-one knows how the true story will end.

The analogy is not perfect, nor was the depreciation analogy involving the sun and the moon in Chapter Four. The value of an analogy is in the light which it casts on a phenomenon, and different analogies cast their light from different directions. The tide analogy was chosen partly because, unlike the ancient Phoenicians, we (in Britain) have been familiar with the tide since the reign of our Anglo-Saxon King Knut. Every weekend throughout the summer thousands of yachtsmen study their tide tables meticulously before setting sail and they disdain the minority of their fellows who omit that precaution, as foolhardy and a menace to other craft. Yet when they return to their offices on Monday morning, they spend billions of pounds with either no regard for inflation or at best no more than some very hazy notions about its effects on their decisions.

As mentioned in Chapter Three, there was not much inflation in Britain in peacetime up to the Second World War, but from 1947 to 1997 it averaged 6.1% a year, reaching 24% in 1975. In general the cost of what £1 would buy in 1947 increased by 19 times; or, putting it the other way round, the value of money fell by nearly 95%. That is but a modest annual rate compared with the hundreds of percent a year under which some countries have suffered, but it has seriously distorted the published Historical Cost accounts of thousands of companies throughout those 50 years and those accounts have misled millions of investors and others in consequence. Yet few companies state the extent of the distortion in their annual reports because there is no agreed method for them to do so.

In Chapter Four, some space was devoted to showing why the most common method of calculating depreciation might be having harmful consequences. That was necessary because the accounting profession has been unwilling to concede that there is a substantial problem. In this chapter the situation is different. Accountants have been openly worried about 'inflation accounting', as it is called, ever since 1949, when the

Institute of Chartered Accountants in England and Wales started publishing recommendations on the subject[3], but there have been two schools of thought among accountants as to the nature of inflation and and its effects on companies and their profits. The minor school spoke of the 'purchasing power' of money and in 1974 devised the 'Constant Purchasing Power' method of accounting for inflation, or CPPA for short. The other, dominant school spoke of the 'current cost' of replacing fixed assets and stock-in-trade and devised a set of rules, called 'Current Cost Accounting', or CCA for short.

As mentioned in Chapter One, CCA was imposed on a reluctant industry in Britain for nine years and then withdrawn in 1988. Had the essentially technological objections to CCA, which had been brushed aside when it was imposed, been admitted when it was withdrawn, that would have ended the episode with some hope that the problems of inflation accounting, which had by no means disappeared, would now be addressed with an open mind. But they were not. The perpetrators of CCA continue to parade their handbook as an authoritative reference work and nothing has been offered in its place. CCA therefore remains as an intellectual barrier against the adoption of an adequate system of inflation accounting in the interest of deriving prosperity from technology.

THE REPLACEMENT FALLACY

CCA was recommended by the Sandilands Committee[4], whose report in 1975 denied that general price changes could be quantified, notwithstanding that they had been quantified every month for the previous 65 years. There is no need to argue the point again because another committee firmly reasserted in 1976 that they could be quantified by the application of the Retail Prices Index[5], but it was already too late to stop the profession from wandering into the CCA quagmire.

The notion on which the Sandilands Committee fastened was that the provision for the depreciation of fixed assets should be based on the costs of replacing them at the ends of their economic lives instead of on their historical acquisition costs, and that the annual depreciation of a fixed asset should be based on the current cost of replacing the value 'consumed' during that year. If one has the fixed belief that money is the absolute measure of value and one has no insight into the causes of depreciation, the idea is probably seductive because fixed assets often do have to be replaced, in a sense, at the ends of their service lives and the replacements often cost much more than the original assets. But the replacements may bear little or no resemblance to the original assets and quite often assets are not replaced at all. Moreover the 'consumption' metaphor is misleading because it gives the impression that depreciation comes from using the asset whereas, as explained in Chapter Four, the main cause in most cases is external technological progress.

In Chapter Four it was emphasised that technological progress is gradational. Now it must be emphasised that it is also cumulatively extensive and often radical. In the case of electricity supply, a generating station such as was described in Chapter Two will probably be replaced, in some sense, by a station or stations which also generate(s) electricity, but the essence of the resemblance is no deeper than that. The replacement may burn a different fuel by a different method or not consume any fuel. If it does consume fuel, it may not raise steam. So how is one to calculate the cost of replacing such an asset?

In the 25 years up to 1979 (when CCA was imposed), economies of scale and better methods of manufacture and construction had reduced the real capital cost per kilowatt of capacity for any given design of station; but on the other hand the improvements described in Chapter Four (page 117) had increased the capital cost which it might be economical to incur in order to reduce the subsequent operating costs. The possible range, arising from those opposing trends, was from about £180 to £550 or so per kilowatt, and within that 3:1 range the best choice depended – and of course still depends at whatever time this book is being read – on the prospective fuel prices and the intended outputs (in relation to their

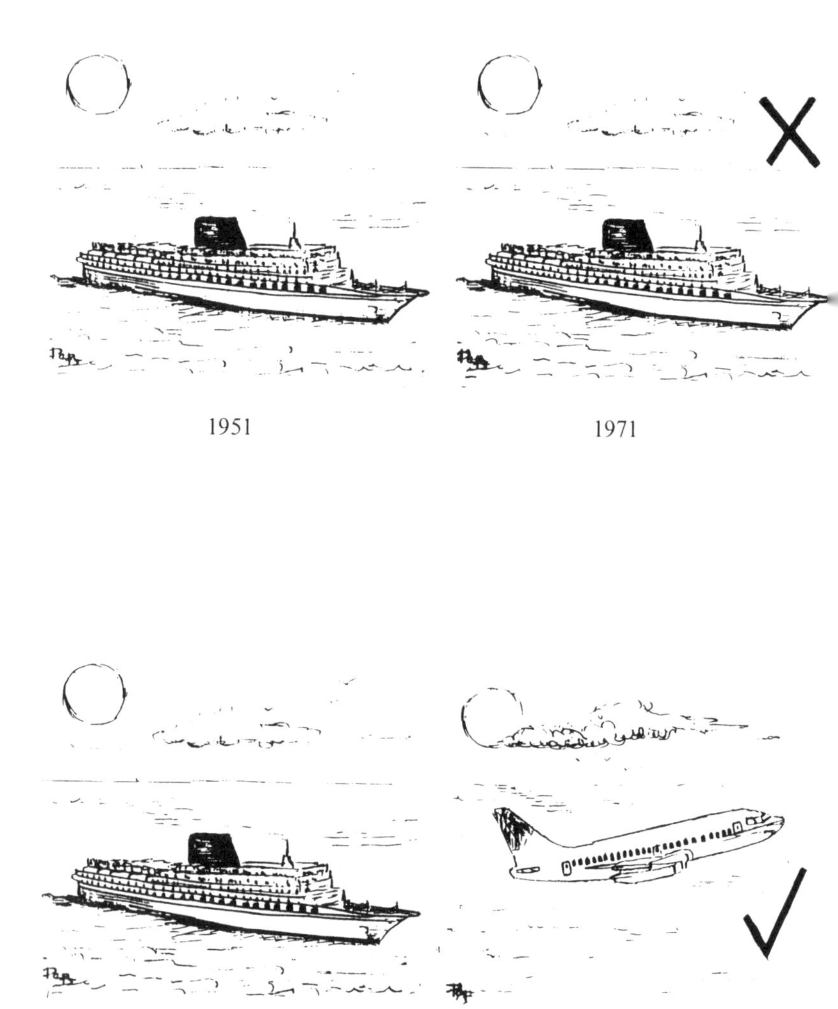

... the replacements may bear little or no resemblance to the original assets ...

output capacities) of the new stations during their service lives, among other factors. Calculating when it would be economical to take old stations out of service, and designing the new ones, were and are the major occupations of technologists in that field.

Some background papers, issued in conjunction with the CCA manuals, apparently perceiving the futility of tying accounts to obsolete technology, devoted a couple of pages to 'the impact of technological progress on gross current costs [of replacing fixed assets]' and stated that 'the effect of technological advance on the current value of assets should be reflected in the accounts...'. But the calculations would be many times more subtle and complicated than those papers envisaged, and they would require a detailed knowledge of the technology involved, together with a decade or two of responsible experience. They would be impossibly expensive to perform (with any pretension of accuracy) every year for every fixed asset, and the notion that they could be effectively audited was and is ludicrous.

Generating stations are by no means an extreme case. Shipping (as mentioned in Chapter Four), petroleum refining, the chemical industry, gas supply, the pharmaceutical industry, spinning and weaving, food processing, printing, transport, telecommunications – the list is almost endless – have all undergone extensive technological progress in the last 50 years and there is no reason to suppose that it will not continue. The considerations affecting the replacement (or not, as the case may be) of fixed assets in those industries are as complicated as they are in electricity supply.

In the case of generating stations, at least the energy is still sent out as electricity, but in transport the changes are often more radical. For international passenger transport, aeroplanes have almost wholly superseded ships, and in Britain roads have largely superseded railways for freight transport. In many cases coal is no longer transported at all but is burned in generating stations at the coal mines and the energy is transmitted as electricity to where the coal was previously burned. So, if the intention of those first manuals were taken seriously, CCA would require the gross replacement cost (the first CCA step in calculating the value supposedly consumed) of a railway to take account of the cost of generating electricity! There is no limit to the complexity of the calculations or to the depth of the quagmire into which accountants strayed when they adopted the replacement proposition.

It has been argued *extempore* that rail transport and electricity supply are different businesses, and that the calculations need not go outside the business which owns the assets. But electricity supply undertakings would build and operate their own railways to transport their fuel if it were economical to do so – indeed they have done so in some countries. The extent of such vertical integration (as it is called) is not governed by any natural law which says that, for example, mining, transport and electricity supply are different businesses and that transport undertakings own their own tracks (in road transport they do not) or that electricity supply undertakings own their own boilers and generators (sometimes they do not) but not their own mines or railways (sometimes they own both). Business boundaries are determined by chance and circumstance – sometimes even by the whims of industrialists. So there is no way out of the quagmire in that direction.

Perhaps for such reasons, yet another CCA committee, after some agonising, simply ignored the whole problem of technological progress, and reverted to relying on special indices which set out to exclude it[6]. But that is tantamount to recommending that the cost-price of electricity today should depend in part on the capital cost of reproducing at current prices generating stations whose designs were superseded 30 years ago and went out of regular production soon afterwards – the very absurdity which the authors of the earlier background papers had sought to avoid.

The mistake was that 'replace' was not properly understood and defined in the first place. If the inherent ambiguity of the term, which was repeatedly pointed out during the 1970s[7], had not been swept under the carpet, CCA would not have been born or at least not imposed as an accounting standard.

A problem for small traders

However some readers, while ready to grant that the foregoing considerations are possibly relevant to large companies and sophisticated applications of technology, may be nevertheless reserving the thought that small traders are also affected by inflation and need to be able to replace their fixed assets at the ever higher prices which inflation undoubtedly causes. If so they may be reassured to find – if they read on – that, with two provisos, the adjustments to normal published Historical Cost accounts, proposed in the next section of this chapter, would enable companies of any size to replace their assets at those higher prices if they so wished, without dipping into their profits. The provisos are: (1) that the companies can provide for such replacements when there is no inflation, and (2) that the prices of the replacements shall have risen by no more than the general price index during the service lives of the original assets.

'Ah! but what about when the replacement prices do rise by more than the rate of general inflation?' those readers may ask. 'Is a business to go bankrupt just because its suppliers of fixed assets raise their prices by more than the RPI, which is determined by the prices of food and other things having no relevance to that business?' The answer to that sort of question is: 'We are talking about keeping accounts, which is only a small, though vital, part of running a business and keeping it solvent.' There is nothing to prevent a prosperous business, large or small, from putting aside part of its profits in advance to acquire fixed assets at higher real prices next time round before declaring a dividend, but the accounting question is: should the higher real costs of the new assets be accounted as part of the cost of running the business before the new assets are acquired or after they come into service?

For example, suppose Mr Wall, a builders' merchant, buys a small van with a petrol engine for £10,000. Eight years later he wants to scrap it and buy a new one. In that time the general index has risen by 50% (i.e. by an average of 5.2% a year), so Mr Wall is able to put up £15,000 from his depreciation fund using the accounting method proposed in the next section. But he finds to his dismay that the new van will cost £20,000, partly because manufacturing wage rates have risen faster than the general index and partly because the new van will have a diesel engine, which is inherently more expensive to manufacture but will burn a cheaper fuel in smaller quantities for the next eight years. To ensure that the crucial issue cannot be evaded, let us stipulate that the old design of petrol van is now obsolete and no longer manufactured, so Mr Wall must spend £20,000 or find another way of transporting his materials or go out of business. He cannot escape from the consequences of technological progress just by being small. The question is whether the extra £5,000 is part of Mr Wall's business expenses before or after the date of the replacement.

That was the very question which was raised in general terms in the aforesaid CCA background papers and over which the third committee agonised. Its decision to ignore technological progress meant that the published price indices for CCA (in the words of the publication) 'measure the change in price of products of constant specification, so far as possible' and thus 'provide an estimate of what an existing asset would cost to replace if a new asset of the same specification were still available: not the cost of replacing it by an asset of modern design with the same productive capacity.' So if Mr Wall adopts CCA and can manage to do all the calculations correctly, his funds for the replacement will include an indexed allowance for the higher wage rates in making a petrol van but not the additional cost of a diesel van, although it is the only van available. Mr Wall, as a pragmatic business man with no great interest in the niceties of accounting, may conclude that half a loaf is better than no bread and opt for CCA, but this is a cautionary tale.

In passing, that elementary example gives just a glimpse of the appalling problems of collecting the data and composing indices for all the different kinds of assets used in a modern industrial society, correcting the indices to eliminate the effects of changes in the specifications and then applying them. How the compilers of such indices are supposed to calculate the current cost of assets which went out of production many years previously is not explained, but the little phrase 'so far as possible' protects them from criticism if they cannot. That sort of complexity and unreliability contributed to the eventual withdrawal of CCA, but it is not merely impossibly complicated; it is also theoretically absurd, even for Mr Wall.

To resume the cautionary tale, let us say that in the next eight years inflation drops to one per cent a year, amounting to 8.3% over the whole period, and there are no further improvements in the design of vans, but in that time the manufacturers gradually invest in sophisticated robots, which halve the real cost of making the vans. The published index numbers for CCA follow that trend. So at his next replacement Mr Wall finds that his CCA fund for replacing his van amounts to one half of £20,000 + 8.3%, that is £10,830. But his bank, from which on this occasion he borrowed to buy the van, expects to be repaid £20,000 – plus interest of course, but that is not the problem. Mr Wall is left with a debt of £9,170 and no asset against which to set it in his balance sheet. CCA, which seemed to be tailor-made for him to tackle specific price increases in his own neck of the woods, leaves him badly unprotected when real prices drop in that neck. And we must not forget that one of the aims of technologists, which they very often achieve, is precisely that – to reduce the real cost of producing or providing a good or service without changing its specification.

That stark situation did not arise (or arose very seldom) in practice because large companies were not expected to rely on CCA; they kept the old-fashioned Historical Cost accounts, with no allowance for inflation, as well, so they could use their judgement on which accounts to use for any particular decision. And small companies were not obliged or encouraged to adopt CCA, which was just as well since people like Mr Wall might have been innocent enough to believe the figures in their CCA accounts, and been led into grave financial difficulties in consequence.

The term 'replacement' has something in common with the term 'energy'. Both are seemingly straightforward concepts which people who have not fully understood them have tried to use as the foundations for elaborate calculations affecting important decisions; but both are really rather elusive concepts and they are too fragile to withstand the weight of such calculations. We need something more robust and more consonant with the realities of technology. To find it we must return to the principle of Historical Cost.

ADJUSTED HISTORICAL COST ACCOUNTING

The accountants' Recommendation N9 on the annual depreciation of fixed assets back in 1945 was severely criticised in Chapter Four, but the definition in N9 of total depreciation over the whole service life of an asset was not criticised. That definition was: 'Depreciation [that is total depreciation] represents that part of the cost of a fixed asset to its owner which is not recoverable when the asset is finally put out of use by him.'

In other words: total depreciation = acquisition cost – final disposal value (if any). The Historical Cost principle is firmly vested in that definition and should not be disturbed, but in 1945 'cost' meant cost in money, which provided the absolute measure of value, at least in Britain. Now the absolute measure of value is 'real value', as described in Chapter Three. So the definition should be amended to read: '*Real* depreciation represents that part of the *real* cost of a fixed asset . . .'.

Unfortunately real value is still an unfamiliar measure and its adoption saddles annual accounting with an awkward choice. Should the costs and values in annual accounts be real costs and real values or money costs and money values? The same choice was encountered in adapting the cost method to cater for inflation (Chapter Three, page 83) but there the balance of convenience was heavily on the side of real values and real costs. In annual accounting the balance is more delicate.

A set of Historical Cost accounts comprises an end-of-year balance sheet and a profit and loss account for the year, both accompanied by their counterparts from the previous year so that the reader does not have to hunt up the previous year's accounts to see whether and in which respects the company's fortunes have risen or fallen since then. The items in the end-of-year balance sheet all represent values on the same date, so there is no difference between the money value of an item and its real value at the general price level at that date, the end of the accounting year. But the other three components are more awkward. If the accounts are to represent real values and real costs, those other three components must also be expressed in real values at the general price level at the end of the year, but the items in the profit and loss account have been calculated from transactions throughout the year. So must the amounts of those transactions all be converted to the end-of-year price level before being added up? And must all the items in last year's balance sheet and last year's profit and loss account also be converted to the new end-of-year price level? And are those converted accounts and balance sheets to be presented as well as their unconverted counterparts, making eight sets of figures for the reader's attention instead of four?

The CPPA school of accountants, whose proposals were overridden by the CCA school, answered all those questions firmly in the affirmative[8]. Had the members of that school succeeded in persuading their fellow accountants to adopt CPPA, as they very nearly did in 1974, their view on that issue of money values versus real values would have been accepted by all and sundry, CPPA would have enabled the concerns which used it to contribute to prosperity as well as they could before inflation became a serious problem, and this chapter would not have been written. But they failed to do that, mainly because real value was an unfamiliar measure but also because CPPA was rather cumbersome, though not nearly as cumbersome as CCA became; moreover the withdrawal of CCA has not been followed by a revival of CPPA, so the field is now open.

The alternative, which was also rejected in favour of CCA but (unlike CPPA) without serious consideration, is to continue keeping the accounts in money values but to adjust the items for inflation in the process of assembling them into the conventional forms, that is to say to make the opposite choice to that which technologists normally make in using the cost method, as described in Chapter Three (page 83). An appropriate name for such accounting is 'Adjusted Historical Cost Accounting', or 'AHCA' for short[9], and its main advantages over CPPA are: (1) its values are money values, which are familiar, (2) only a few adjustments are needed to the traditional Historical Cost accounts and (3) there are only four sets of figures for the reader to absorb, the same number as in ordinary, unadjusted Historical Cost accounts and with almost the same headings. One large disadvantage of money values in the cost method – that the rate of inflation must be forecast on scant information many years ahead – does not apply in annual accounting because the rate for the year of account is known by the time the accounts have to be presented. Let us consider the main features of AHCA.

The adjusted *balance sheet*

Taking the balance sheet first, it contains the net values of the company's assets and liabilities, which are of two kinds, known as 'monetary' and 'non-monetary' assets and liabilities. Monetary assets and liabilities are sums of money which the company is owed by its debtors and which it owes to its creditors and its providers of loan capital, such as outstanding invoices and loans and its debentures and preference shares. All these items are fixed in money, so their money values are not affected by inflation[10]. They need no adjustment in the AHCA end-of-year balance sheet.

The non-monetary assets are the company's physical assets, namely its fixed assets, its stock-in-trade and its work-in-progress. In the Historical Cost accounts, the values of those items are calculated from their acquisition costs and, in the case of the fixed assets, their subsequent depreciation, but ignoring the effects of inflation. Those values must therefore be adjusted for inflation and the most important adjustments are to the net values of the fixed assets. Some fixed assets, and perhaps some stock-in-trade, may also be revalued from time to time to allow for unforeseen changes in their real values, but it is convenient to tackle the effects of general inflation by itself in the first instance.

If a bus in the fleet in Chapter Four is acquired at a cost of £60,000 and it depreciates in its first year by £7,500 with no inflation, then with no inflation its net value at the end of the year is £52,500. It follows that that is its end-of-year net real value at the start-of-year general price level, whatever the rate of inflation. So if inflation runs at the rate of, say, 8% in that year, then at the end of the year its net money value and its net real value at the end-of-year price level are both £52,500 + 8% = £56,700[11].

The adjusted *profit and loss account*

The next question concerns the profit and loss account. It is: how much have the buses depreciated, or in other words how much have their values fallen? That may sound like a circular question because we have just calculated their net values from their depreciation, so how can we now calculate their depreciation from their net values? But it is not circular because what we calculated was their adjusted net values from their real deprecation and what we want now is their depreciation in money. But 'depreciation in money' is a clumsy phrase; a clearer description is 'apparent depreciation', which we can define as simply the difference between the money values at the start and end of the year.

The apparent depreciation, in that sense, of a bus in its first year of service is £60,000 – £56,700 = £5,300 and the apparent depreciation of the fleet is £17.00 million – £15.12 million = £1.88 million. That is therefore the figure which goes into the adjusted profit and loss account, under the heading 'Apparent Depreciation' to make it clear to the readers that the figure is not the unadjusted depreciation figure in a Historical Cost profit and loss account, which ignores inflation, nor is it the real depreciation in a Current Purchasing Power profit and loss account.

The stock-in-trade and work-in-progress are treated in the same ways in both the adjusted end-of-year balance sheet and the adjusted profit and loss account, except that there is normally no real depreciation in their case. So, except in so far as they are physically reduced in quantity during the year, their adjusted values in the balance sheet will tend to increase, leading to their apparent appreciation (or, if you are mathematically minded, negative apparent depreciation) in the adjusted profit and loss account, but that tendency is not cumulative because those assets are held for the purpose of consumption or sale.

Under rapid inflation, depreciating fixed assets with long economic lives can also undergo apparent appreciation in the early years of their lives, but that tendency must be reversed in their later years because, although they are not held for consumption or sale, their net values must fall to zero (or to their final disposal values) in the end. Many home owners understand the meaning of apparent appreciation due to inflation, apart from the effects of housing shortages and land shortages on the real values of their properties. In their case the eventual fall in value is often so long delayed that it comes to be disregarded, but even houses, though traditionally safe as investments, have finite economic lives and must be scrapped in the end.

Accountants who favoured CPPA have condemned those figures as unreal and misleading, but although they are unreal they are not necessarily misleading. If you are sailing directly against the tide and your high-tech speedometer (which performs the same function as the old sailors' log and line) reads 6.4 kilometres an hour, you can react in one of two ways. You can adjust the reading immediately for the speed of the tide or you can enter 6.4 in your log under 'apparent speed', then continue with your readings of wind velocity, compass course, etc. and make the adjustment for the speed of the tide at a later stage in your calculations – when it suits you. In an AHCA balance sheet we adjust the readings as they come in, but in the profit and loss account we are always sailing directly against the tide of inflation and we choose the second way. It suits us to make a single adjustment for inflation at the end of the calculation rather than item by item.

The equity adjustment for inflation

The adjustment, when it comes, is simple. We first subtract the total value of all the liabilities from the total value of all the assets in our balance sheet. The difference represents what the ordinary shareholders have invested directly in the business (rather than indirectly by buying shares from existing shareholders) plus the gains which the business has made but not distributed in ordinary dividends. Let us label it 'the equity capital'[12]. It is not the same as the equity valuation of the business on the Stock Exchange because that depends on the inevitably speculative views of the buyers and sellers of ordinary shares, and it is moreover a fictional valuation in as much as it is normally the product of the total number of ordinary shares and the price at which only a few shares change hands from time to time. The equity capital, as defined, is the outcome of a strictly non-speculative, non-fictional set of calculations, derived as far as possible from the Historical Cost principle.

In deference to modern accounting practice, the equity capital can be calculated in two steps. Gone are the days when annual accounts were presented as long lists of credit items opposite long lists of debit items, adding up to the same totals, to the mystification of most non-accounting recipients. Nowadays they come as mixtures of additions and subtractions under a few portmanteau headings, with nearly all the details removed into pages and pages of Notes to the Accounts, which are supposed to make the whole package easier to absorb. One of the portmanteau headings is 'Stockholders Funds'[13] and its value is obtained by subtracting all the liabilities except the preference shares from the total assets. Those funds are then sub-divided under various headings, but they do not concern us. All we need for our second step is to subtract the values of the preference shares from the stockholders' funds to obtain the equity capital.

Returning to the adjustment, its purpose is to guard the real value of the equity capital against the depredations of inflation. However the equity capital at the end of any year normally includes the dividends which are declared in the accounts for that year but not yet distributed. The equity capital at the start of the current year is therefore the equity capital at the end of the previous year minus the dividends declared then. The adjustment is the product of that net figure and the rate of inflation in the current year[14] and it is a supplementary debit item in the adjusted profit and loss account. If the equity capital at the end of last year was £1 million, from which £50,000 was declared in dividends, and inflation this year ran at 8%, the adjustment would be 8% of £(1,000,000 – 50,000) = £76,000. A convenient label for the adjustment is the 'equity adjustment for inflation', or 'equity adjustment' for short. Its effect is to reduce the profit in the current year and increase the equity capital at the start of the next year by the same amount.

Those few adjustments are all that are required to turn a set of conventional Historical Cost accounts into a valid set of Adjusted Historical Cost Accounts. On the next page, the main headings are compared with those of a corresponding set of unadjusted Historical Cost accounts. The size and complexity of a business inevitably affects the layout and headings of its accounts, as well as the figures under them, so those headings are typical, rather than definitive, but in conjunction with the notes at the foot of the page they encapsulate the principles of AHCA.

The question then arises: should last year's balance sheet and last year's profit and loss account be simply copied from last year's annual report or should they be converted to allow for inflation between the two sets of accounts?[15] To a technologist, or anyone else who is not put off by a few columns of figures, such a question is on a par with whether calorific values should be net or gross, as discussed in Chapter Two. It hardly matters as long as the choice is clearly stated and the necessary data for the conversion or reversion are appended, so that anyone who wants the other figures can pull out his or her pocket calculator and work them out in a few minutes. However, it would be convenient if everyone made the same choice.

The simpler of the two options is not to convert the figures from last year's accounts, but rather to use the said Notes to

HISTORICAL COST ACCOUNTS	***ADJUSTED* HISTORICAL COST ACCOUNTS**	**Notes**
	BALANCE SHEETS	
Fixed Assets	*Adjusted* Fixed Assets	1
Current Assets	Current Assets	
Stock-in-trade	*Adjusted* stock-in-trade	1
Work-in-progress	*Adjusted* work-in-progress	1
Debtors	Debtors	
Bank balances and cash	Bank balances and cash	
Current Liabilities	Current Liabilities	
Creditors	Creditors	
Stockholders' Funds	Stockholders' Funds	2
	Preference Share holdings	
	Equity Capital	3
	PROFIT AND LOSS ACCOUNTS	
Turnover	Turnover	
Cost of Sales	*Apparent* Cost of Sales	4
Profit Before Taxation	*Apparent* Profit Before Taxation	5
	Equity Adjustment for Inflation	6
	Adjusted Profit Before Taxation	7
Taxation	Taxation	
Profit After Taxation	*Adjusted* Profit After Taxation	

Notes

1) At end-of-year price level
2) As calculated from the foregoing adjusted values
3) Stockholders' Funds less Preference Share holdings
4) Including apparent depreciation of fixed assets, stock-in-trade and work-in-progress
5) As calculated from Apparent Cost of Sales
6) Equity Capital less declared dividends, multiplied by the rate of inflation[16]
7) Apparent Profit less Equity Adjustment for Inflation

explain in detail, item by item, how much of the difference between this year's figures and last year's figures was due to inflation and how much to the expansion or contraction of the business, unforeseen events and so on. The one suggestion to be sternly resisted is that eight sets of figures should be presented in the main pages of the accounts instead of four. Those eight sets of figures gave the kiss of death to Constant Purchasing Power Accounting and they would do the same to Adjusted Historical Cost Accounting if they were forced on to the main pages.

The wedge formula in AHCA

The principal adjustment in AHCA is the equity adjustment. Its effect is substantially to reduce the profit in the Historical Cost accounts. It is the product of the rate of inflation during the year of account and the net equity capital at the start of the year. The rate of inflation is measured externally, but the net equity capital is obtained by an internal calculation, which is therefore critically important. That calculation depends on the net value of the fixed assets in the balance sheet at the end of the previous year and hence on the method of calculating their depreciation in that year and in earlier years.

That is therefore the point at which the penultimate section in Chapter Four, 'The Foreseeable Deprival Value', bears on inflation accounting. It is possible to contemplate a form of AHCA in which some other depreciation method, say the straight-line method or the declining-balance method at the rate of 25% a year, continued to be used. The net values of the fixed assets in the balance sheet would then be no more than the means of carrying forward some totals from one year to the next, with no particular significance attaching to them. For example, for the static fleet of buses the total net value brought forward from the previous year would be £25.5 million under the straight-line method but only £11.9 million under the declining-balance method – a difference of £13.6 million. Consequently, with inflation running at 8%, the equity adjustment for the fleet (assuming it to be financed wholly by equity capital) would be 8% of £13.6 million = £1.1 million less under the latter method than under the former method – but no particular significance would attach to either figure.

It ought not to be acceptable to have figures in annual accounts to which no particular significance can be attached, though they are tolerated in Historical Cost accounts when there is no inflation, as we have seen in Chapter Four. That may have been an additional objection to CPPA when it was rejected in favour of CCA, in as much as CPPA is as sensitive as AHCA to the method of calculating depreciation. But, if so, the objection was misdirected; that particular fault lay not in the inflation accounting method but in the depreciation method.

The depreciation method affects the profit or loss calculation even when there is no inflation, but inflation produces the additional harm that errors in calculating the equity capital are cumulative. For example, the estimated foreseeable deprival value of the static fleet of buses brought forward from the previous year is £17.0 million, which is £8.5 million less than the net value with straight-line depreciation. Straight-line depreciation would therefore make the equity adjustment 8% of £8.5 million = £0.7 million too large and the adjusted profit would be £0.7 million too small. The business would therefore start the next year with its equity capital overstated by £8.5 million + £0.7 million = £9.2 million. If inflation continued at the same rate, that overstatement would rise to £9.9 million next year, £10.7 million the year after and so on. On the other hand the reducing-balance method would understate the equity capital by £3.7 million at the end of the current year, £4.0 million at the end of next year and so on.

Those errors would be permanent because, although inflation might slow down in future years, it will not reverse. Indeed any government would try to take counter-measures if it showed signs of doing so. In other words, AHCA or CPPA, or any other form of inflation accounting which recognises the nature of general inflation, will be permanently flawed unless it is based on the wedge formula for the annual depreciation of most of the non-wasting fixed assets – or of course on a refinement along the second branch of the depreciation theory described in Chapter Four.

Replacement of fixed assets under AHCA

It is time to redeem the promise in the previous section (page 163) that small traders would be able to replace their fixed assets at the higher prices caused by inflation, if they so wished, without dipping into their profits, provided that: (1) they were able to do so when there was no inflation, and (2) the actual prices of the replacements under inflation had risen by no more than the general price index during the service lives of the assets. Where can Mr Wall, the builders' merchant, using AHCA, find £15,000 to replace his van, which cost £10,000 eight years earlier, that is to say with inflation running at an average rate of 5.2% a year, in the absence of other extraneous influences? It has been a vexed question throughout the history of inflation accounting.

The answer depends on where he found the £10,000 for his original van and how he would go about replacing it in the absence of inflation. If he borrowed the money from his bank at a normal money rate of interest, he should use the successive depreciation provisions, as calculated by the wedge formula, to repay the loan. The interest would of course be an additional charge. With no inflation, those depreciation provisions would be £2,500, £2,143, ... £357, £0[17], which add up to £10,000, so he would be left with no debt at the end of seven years. When he scrapped his old van a year later, he would borrow another £10,000 for the new van. Now with inflation at 5.2% a year, the depreciation provisions under AHCA would be £2,110, £1,960, ... £0[18], which again add up to £10,000, so he would again have no debt after seven years. The only apparent difference is that he would have to borrow £15,000 for the new van instead of £10,000, but that is not a real difference because the real value of £15,000 then would be £10,000 at the general price level eight years earlier.

If, on the other hand, Mr Wall wished to use his equity capital to buy the original van, he would have a practical problem which has nothing to do with inflation or the depreciation method and is seldom analysed, but which he would be wise to understand before deciding on that method of finance. The problem is that he would have to be able to invest the successive depreciation provisions as equity capital, either in his own business or elsewhere, as profitably as the rest of his equity capital, and be able to turn those investments into cash when the time came to replace the van. If he left those provisions idle or lent them at a normal money rate of interest, his equity capital base would shrink, year by year, irrespective of inflation, and – other things being equal – his profits would shrink with it. Depending on the complexity of his operations and his or his accountant's acumen, he might or he might not perceive the cause of the shrinkage, but it is a business maxim that all the capital must be put to work if a business is to thrive.

Borrowing money for the intermittent acquisition of fixed assets also has its problems, irrespective of inflation, but this book is not a manual on business finance methods. Provided the business in which Mr Wall temporarily invests his successive depreciation provisions uses AHCA, the equity capital represented by those investments will automatically attract equity adjustments for inflation. There will then be three sources of funds for the replacement, namely the depreciation provisions themselves, the equity adjustments on the decreasing net value of the asset in service and the equity adjustments on the accumulating provisions when invested as equity capital.

The calculation on the next page shows how those three sources can be combined to build up a replacement fund of precisely £15,000 at the end of eight years. For greater realism, it does not even assume that inflation is running at a constant annual rate, only that the general price index, which is taken to be 100.0 when the original van was acquired, rises to precisely 150.0 eight years later. Incidentally the outcome of the calculation is not a special feature of AHCA: the same result would be obtained under CPPA.

Perhaps it appears that it would be simpler to calculate the real depreciation of the van in the successive years, instead of

HOW MR. WALL'S REPLACEMENT FUND BUILDS UP

(1)*	(2)	(3)	(4)	(5)	(6)	(7)	(8)	(9)
		%	£	£	£	£	£	£
0	100.0		10,000					0
		5.20		2,110	520	0	2,630	
1	105.2		7,890					2,630
		5.23		1,960	413	138	2,511	
2	110.7		5,930					5,141
		6.59		1,716	391	339	2,446	
3	118.0		4,214					7,586
		3.47		1,598	146	263	2,007	
4	122.1		2,616					9,594
		4.83		1,245	126	463	1,834	
5	128.0		1,371					11,428
		9.38		871	129	1,072	2,072	
6	140.0		500					13,500
		0		500	0	0	500	
7	140.0		0					14,000
		7.14		0	0	1,000	1,000	
8	150.0		0					15,000

* **Key to columns**
1) Years from acquisition of asset in service.
2) General price index.
3) Inflation rate from (2).
4) *Adjusted* net value by AHCA formula.
5) *Apparent* depreciation from (4), by difference.
6) Equity adjustment on net value = rate in (3) × prior value in (4).
7) Equity adjustment on replacement fund = rate in (3) × prior value in (9).
8) Annual addition to replacement fund = (5) + (6) + (7).
9) Replacement fund = previous value + (8).

going through the rigmarole of imagining the successive provisions for its apparent depreciation to be invested as equity capital, which then attracts a series of equity adjustments. Calculating the real depreciation is the CPPA approach and it leads to the same answer in the end if the calculation is carefully performed, but it is not simpler and it tends to conceal the practical difficulties of providing for the replacement of a fixed asset at the end of its service life. Those successive investments are not imaginary; they are real investments and they incur the real problems already mentioned. It is sometimes very difficult to withdraw equity capital from one part of a business in order to invest it in another part, and even more difficult to withdraw it from one business in order to invest it in another business.

The crucial point is that the equity adjustments are not made to the depreciation provisions as such, but only to the equity capital when they are invested. If they are not invested as equity capital, they will not attract any equity adjustments. If Mr Wall prefers to lend them at a money rate of interest because he thinks that will be safer, he must realise that he is taking some equity capital out of his business every year and that may reduce his profits. He cannot have his cake and eat it.

However Mr Wall can eat any part of his cake and still have the rest of it. For example, suppose he puts up 30% of the cost of his £10,000 van from his equity capital and borrows 70% from his bank. Then 70% of the successive provisions for apparent depreciation (column 5 in the calculation) will add up to £7,000, which means that they will be enough to repay the bank loan in annual instalments. And the remaining 30% of those provisions, duly invested as equity capital, will build up to a fund of 30% of the replacement cost, that is £4,500. That kind of financing is very common, under the name of 'geared' financing. Whether it is to Mr Wall's advantage to have the practical problems of borrowing and those of finding equity capital for his fixed assets, rather than one set of problems or the other, is for him to decide in the light of current interest rates, his opinion of the future rate of inflation and so on, but his AHCA accounts will cope with the effects of inflation whatever he decides.

In regard to that last assertion, however, we should perhaps just remind ourselves that we are talking about the comparatively modest rates of general inflation mentioned at the start of this chapter (page 157). Reverting to the marine analogy, we are sailing in tidal waters, not shooting the rapids in a canoe on a river through a mountain gorge at the end of the rainy season. AHCA is not intended to cater for inflation running at several hundred per cent a year – and nor of course was CPPA or CCA.

To conclude this sub-section, the fact that AHCA would enable companies and undertakings, large or small, to replace their fixed assets under certain conditions without dipping into their profits, does not weaken the argument that replacement is not the proper purpose of providing for depreciation. In most circumstances providing for replacement is a blinkered pursuit. It misses the new opportunities which technological progress opens up between the acquisition of fixed assets and the expiry of their service lives.

Resources for measuring technological progress

In Chapter 4 (page 131) the possibility was mooted of refining the rough and ready estimates of the economic lives of assets which suffice for the discounting calculations, and it was stated that a possible source of funds and of some of the relevant expertise would emerge in the next chapter, that is now the present chapter. Those sources in Britain are the funds and staff of the part of the Office for National Statistics which, at the time of writing, are still engaged in collecting data and compiling and publishing hundreds of price index numbers for CCA (see page 162 and Footnote 6), although CCA was withdrawn ten years ago.

Up to now they have been trying to 'measure the change in price of products of constant specification, so far as possible' and thus 'provide an estimate of what an existing asset would cost to replace if a new asset of the same specification were still available; not the cost of replacing it by an asset of modern design with the same productive capacity'. They could now be requested to abandon those quests and try instead to measure the changes in specification, design and real prices (as adjusted by the RPI) of classes of assets with the same productive capacity, in other words to measure the rates of technological progress in specific sectors of the economy. Much of the data which they have been collecting for the former purpose would be relevant to the latter purpose and the statistical techniques would be similar, and in some respects precisely the same.

Such a turn round, competently directed, would lead not only to better estimates of the economic lives of fixed assets but also to better estimates of how those assets depreciate annually under different operating conditions. Both would improve the quality of the annual accounts of companies and undertakings which acquire those assets.

REVALUATIONS OF PHYSICAL ASSETS

In Chapter Four, the essence of the theory of depreciation was that annual depreciation can be foreseen, but that does not mean that it can be perfectly foreseen. Two hundred years ago, Robert Burns mused that:

> The best laid schemes o' mice an' men
> Gang aft agley,
> An' lea's us nought but grief an' pain,
> For promis'd joy![19]

and the schemes of technologists are not exceptions, although their consequences are seldom as totally disastrous as that poem avers. Sometimes the benefits of applications of technology are better than was calculated at their inception. An accounting method must accommodate either outcome.

Sometimes unforeseen events merely curtail or extend the best estimate of the remainder of the economic life of an asset without affecting its present real net value enough to warrant altering the balance sheet for the current year – bearing in mind that the present net value is always an estimate with an inherent degree of uncertainty in its composition. In other cases the present real net value is perceived to be substantially greater or less than was expected, and then the accounts must be amended.

Consider first the procedure when there is no inflation. Suppose a building, costing £10 million, has an expected economic life, when it is constructed, of 50 years, after which it will still be usable but probably something newer will be preferred for sound economic reasons. In the absence of any data on which to base a departure from the wedge formula, its net value after, say, eight years will be £7.03 million. But now suppose that some event, such as a change in the planning laws, is found, on revaluation, to have raised the price at which the building could be sold to £12 million. That is then the net value of the building in the balance sheet and, if its economic life still has 42 more years to run, the cost of continuing to use it next year instead of selling it now should include a charge for depreciation of £571,000 instead of £335,000.[20]

The increase in value from £7.03 million to £12 million, that is £4.97 million, is an 'extraordinary gain' and as such it is excluded from the calculation of the profit because it did not arise in the normal course of business and cannot be expected to recur. In effect it goes straight into the stockholders' funds[21].

Now let us suppose that inflation runs at 5.2% a year throughout the eight years. Then the net value after eight years will be £7.03 million + 50% = £10.55 million before the revaluation and £12 million + 50% = £18 million afterwards. The extraordinary gain is the difference, that is £7.45 million, and it is the real gain at the general price level at the end of the eighth year. It can go straight into the stockholders' funds, just as did the £4.97 million in the absence of inflation. It has no effect on the profit in the current year and no effect on the equity adjustment.

Sometimes it is not obvious whether a particular gain (or loss) should be classified as part of the profit or as extraordinary, and professional accountants have a legitimate claim to be better qualified to judge the question than non-accountants. This section poses no challenge to that claim. When the issue has been decided, the AHCA rules are that the adjusted profit is equal to the apparent profit (excluding extraordinary gains) minus the equity adjustment, as calculated from the net equity capital at the start of the year and the rate of inflation during the year; and the extraordinary gains are not affected by the adjustment. Of course an extraordinary gain can be negative, as can the profit, but the equity adjustment is always positive (because inflation never reverses) and it is always subtracted from the apparent profit.

TAXATION

Profits are one of the forms of unearned income. Another is the interest on loans. The effects of inflation on them are different, but the tax rules raise an issue which is common to both and is easier to consider first in relation to the interest on loans.

As explained in Chapter Three (Appendix 2), inflation reduces the effective or real rate of interest on a loan if the money rate is fixed. The reduction before tax is slightly more than the rate of inflation. But an important source of funds for loans is the savings of private individuals, directly or via their ownership of shares, and they are generally liable for income tax. Let us consider the effect of income tax on the interest from a direct loan by a private individual to a company, which uses the loan to invest in fixed assets for an application of technology.

Any tax must create a gap between the commercial cost of the capital, which is what the borrower pays at the gross rate before tax, and the resource cost, which is what the lender receives at the net rate after tax. With no inflation, the gap is just the rate of tax, say 23% for a basic rate taxpayer, so a gross interest rate of 15% is equivalent to a net rate of 11.55%. The effect of inflation is to widen that gap, unnecessarily and sometimes very harmfully, because the tax is charged on the money rate.

For example, inflation at 5% reduces the real gross rate from 15% to 9.52%, a reduction of a little more than a third, but it reduces the net rate after tax from 11.55% to 6.24%[22], that is by nearly one half. And inflation at 10% reduces the gross rate to 4.55%, a reduction of just over two-thirds, whereas the net rate becomes 1.41%, a reduction of nearly nine-tenths; or looking at it another way, the borrower has to pay more than three times as much as the lender receives because the real rate of tax is (4.55 – 1.41)/4.55 = .69 = 69%.

If you were liable to tax at 23% on your marginal income and you postponed buying a new car, costing £10,000, for a year in order to make such a loan, the cost to the borrower would be £1,500, of which £345 would be deducted for your income tax, leaving you with £11,165. But – other things being equal – the price of the car would have risen to £11,000 meanwhile, so your net gain for waiting a year would be just £165, which is equivalent to £150 at the start-of-year price level. And if you accepted a gross rate of only 12%, you would find yourself £76 short of the price of the car when you came to buy it. In the latter case certainly and in the former case probably, you would be well advised to buy your new car at the outset, thus depriving the borrower of some funds for a probably beneficial application of technology. If there were a million people in that situation, the denial would amount to £10 billion.

For higher-rate taxpayers, the inducement to save is even smaller. With inflation at 10%, taxpayers in the 40% band need a gross interest rate of 16.7% just to maintain the real value of their savings. That corresponds to a real rate for the borrower of 6.1%. But if the rate of inflation falls to 5%, the lenders' real rate rises from zero to 4.8%, while the borrower's real rate falls to 5.4%. The loser in that case is the Treasury, which receives 90% less in real revenue, but none of the real figures appears in any account.

Why there is no general outcry at this practice of taxing unearned incomes at varying real rates up to over 100% (they have been as high as that) is hard to understand. Presumably it has much to do with the general lack of understanding of the workings of inflation and perhaps the fact that most voters still have very little income from loans or investments. The practice reduces our prosperity by raising the real commercial cost of borrowing for applications of technology well above its real resource cost, as determined by the potential lenders' opportunities to spend rather than lend. During inflation, income tax discourages saving much more than it discourages working.

The obvious reform is to tax interest from loans on the real income which it represents. That would produce less revenue, which might be made good by raising the rates of tax on all income or on unearned income. In the latter case, which some people (not including the author) might advocate on putative moral grounds, at least the real tax rate would be known in advance instead of depending on the future rate of inflation, which the lenders and borrowers have to guess when they arrange their loans.

There is one type of loan, namely Index-Linked Gilts, on which the interest is already effectively taxed in that way. The stated interest is paid at a comparatively low rate, but it is on top of full compensation for inflation, and only that interest is taxed, not the compensation for inflation. However those loans are not mainly for investment in applications of technology. How such an inflation-compensation element would be treated if a bank in the private sector offered index-linked accounts to private customers is a matter of conjecture, since the law, at least in Britain, appears to be obscure on that point. Presumably the government has intimated in some way that it would be reluctant to let others do what it does itself, since otherwise the enormous scope for reducing the gap between what the borrower pays and what the lender receives, to the advantage of both parties, would surely have been exploited by now. The important point, which all the parties appear to have overlooked, is that it would also increase our prosperity by increasing the flow of savings for investment in technology.

Turning to the taxation of profits, in 1977 some calculations of the profits and taxes of banks were made for the *Bankers' Magazine* because the form of CCA mooted at that time posed some special problems for banks. In the previous year, when inflation was 15.6%, the four English clearing banks had declared pre-tax profits totalling £700 million and paid £368 million in tax, but their real pre-tax profits totalled only £271 million. In other words they had been taxed at the confiscatory rate of 136% on their real profits. Equivalent calculations for two industrial concerns showed broadly similar results, namely real pre-tax profits much less than the published profits and a real taxation rate of 169% in one case and 97% in the other. The calculations were published in December 1977[23].

If the same conditions were to recur (and who can be confident that they will not?), the published accounts would be just as misleading and the real taxes would still be confiscatory (although the tax rate is lower now). Those published and real pre-tax profits would be £2.1 billion and £800 million respectively at today's general price level.

The taxation of profits has the same main feature as the taxation of interest on loans, namely that the rules disregard inflation, but the rules for profits are also more complicated in that they also disregard the depreciation of fixed assets and introduce writing-down allowances, as described in Chapter Four (page 142), instead. Neither CCA nor CPPA had any effect or potential effect on those rules because neither had anything equivalent to the equity adjustment for inflation, proposed in AHCA.

The general principle in taxing profits is that the tax is calculated from the accounted profit, except in regard to depreciation. It might therefore be argued that the tax rules, which disregard inflation, do so only because the accounts disregard it. If so, under that general principle, AHCA would automatically reduce companies' liability to tax on their profits (or 'corporation tax', as it is called) by the amount of the equity adjustment. However there is little point in pursuing such a hypothetical line of thought because it would be much more effective and less tortured to establish a firm rule that all taxes on income which depends on lending or investing shall be calculated on the real interest or profit.

For administrative convenience, apparent writing-down allowances at a fixed annual percentage could be calculated in the same way as the apparent depreciation earlier in this chapter (page 169)[24]. The real value of the adjusted post-tax profit would then be the same as the post-tax profit calculated by ordinary Historical Cost accounting in the absence of inflation. The calculations would be cumbersome, but not much more so than the present unpublished calculations of the writing-down allowances. However it is difficult to believe that administrative convenience is the real reason for not extending the general principle to include depreciation in the profit calculation for taxation.

REVIEW OF PROGRESS

It is sometimes asserted that the only solution to the problems of inflation accounting is to abolish inflation, but that is like saying that the only way of dealing with injuries from road accidents is to abolish collisions. It might be possible one day to do both, but up to now both problems have proved intractable in societies which use money and build roads. We are not going to abandon roads or money, so we had better make sure we have good hospitals and a good inflation accounting system. If one day we find we have no further use for them, that will be a day of rejoicing and we shall not begrudge the efforts we have put into providing them.

Adjusted Historical Cost Accounting is not a perfect accounting system because there is no such thing. Its claim is that it does what is necessary with the minimum number of changes to the long-established practices of Historical Cost Accounting, to remove their inherent deficiencies and to bring them into accord with the method evolved by technologists for dealing with inflation in designing their projects[25]. However the word 'minimum' applies strictly to the number of changes, not to their magnitudes, which would be substantial.

Charles-Maurice de Talleyrand was reputed to have remarked that: 'War is much too serious a thing to be left to military men'[26]. And of course the same sort of thing has since been said of other professions – of politics and politicians (by bishops), of religion and bishops (by politicians) and so on. In the same vein accountants have often pressed strong advice on technologists. Now the boot is on the other leg, but the truth underlying the badinage is that no branch of knowledge or wisdom can be isolated from the others.

Unfortunately the comments and criticisms of one profession by a member of another are almost always first ignored and then, if that fails, attacked and denigrated, often with scant reference to their content. Technologists are not exceptions to that dreary tendency. But fortunately, in a free society, such comments and criticisms are eventually examined dispassionately.

REFERENCES AND FOOTNOTES

1) Jonathan Swift, *Polite Conversation*. Dialogue 3.
2) However the story has nothing to do with the reference to predicting the tide in Chapter Four.
3) Recommendations N12 on 'Rising price levels in relation to accounts' in 1949 and N15 on 'Accounting in relation to changes in the purchasing power of money' in 1952.
4) Inflation Accounting. 'Report of the Inflation Accounting Committee' (Chairman: F.E.P. Sandilands, OBE), Sept. 1975, Cmnd 6225.
5) 'Guidance Manual on CCA' by the Inflation Accounting Steering Group (Chairman: D Morpeth), Tolley, December 1976.
6) 'Price Index Numbers for CCA', Office for National Statistics, monthly.
7) Explicitly in 'The replacement fallacy in accounting' by the author in *Accountancy*, March 1979, and also in general terms by the author and others in that and other journals (*Management Accounting, The Bankers' Magazine*, etc.) in that and earlier years.
8) Provisional SSAP7, 'Accounting for changes in the purchasing power of money', 1974. Appendix 2 presented the eight sets of figures
9) The principles of AHCA were first proposed and described by the author in the following four articles in 1976-79:

   'The best is the enemy of the good', *Accountancy*, September 1976,

   'Completeness versus universality', *Management Accounting*, April and May 1977,

'Flat-earth accounting', *The Bankers' Magazine*, September 1977,

'The replacement fallacy in accounting', *Accountancy*, March 1979.

The present description includes some modifications and changes of emphasis, including the new name.

10) In the normal course of trading a business discharges its obligations to its preference shareholders by paying them dividends at prescribed rates, which do not vary with inflation or the fortunes of the business. Preference shares are therefore treated as monetary liabilities in the present context, along with the debenture holdings and other loans. Preference shareholders have some latent rights of ownership, but those rights are in abeyance as long as their fixed-rate dividends are paid in time.
11) The general formula is $V = U(1+i)(z-2)/z$, where i is the rate of inflation during the year of account, that is to say $1+i$ is the ratio of the RPI at the end of the year to the RPI at the start of the year. The formula is a straightforward extension of the formula for V which was derived in Chapter Four (page 124), using the same symbols with $F = 0$.
12) It has been referred to as the 'ordinary shareholders interest' or 'equity interest' (notably in SSAP7) but the 'interest' has nothing to do with the interest paid or received when money is borrowed or lent.
13) Often referred to as 'Capital and Reserves'.
14) For accountants. Except that, when new equity capital comes into the business, e.g. by a rights issue, or leaves it during the year, the adjustment for that new or departed capital should be calculated from the inflation between that date and the end of the year.
15) By multiplying every item by the ratio of the respective RPI numbers, i.e. end-of-year to end-of-previous year for the items in last year's balance sheet and middle-of-year to middle-of-previous-year for the items in last year's profit and loss account.
16) Subject to the refinement stated in the footnote 14.
17) From the formula for D in Chapter Four (page 124), with $F = 0$.
18) Using the formula in footnote 11 above.
19) 'To a Mouse (on Turning Her Up in Her Nest With a Plough)', November 1785.

20) £7.03 million × 2/42 = £335,000 but £12 million × 2/42 = £571,000.
21) Under straight-line depreciation, the net value after eight years would be £8.40 million and the extraordinary gain therefore only £3.60 million, instead of £4.97 million, but that is not a valid argument in favour of the straight-line method. On another occasion the value after revaluation might be less than the net value by the wedge formula, in which case the extraordinary loss would be greater under the straight-line method.
22) The formula is: $p = \{r(1-t) - i\}/(1+i)$, where p is the real net rate of interest after tax and t is the tax rate – 23% in the example.
23) 'Three steps to inflation accounting – but one will do', by the author. The banks were Barclays, Lloyds, Midland and Nat West.
24) The formula would be: $V = U(1+i)(1-b)$, where b is the fixed annual percentage (25% for most assets).
25) The fact that the discounting method uses real values, whereas AHCA uses money values, does not make them incompatible.
26) Quoted by Briand to Lloyd George during the First World War.

6

RATIONAL PRICING OF GOODS AND SERVICES

It was stated at the beginning of Chapter Three (page 50) that the use of the cost method is a necessary but not a sufficient condition for deriving prosperity from technology. The essential concomitant is a rational system of prices and methods of charging with the specific purpose of linking the decisions of the designers of projects to the often remote locations where the benefits to the community of those decisions will accrue. Thus in Chapter Two the decision to use copper, rather than silver, for the conductors of a generator[1] must be linked to the copper/silver mine by the prices of copper and silver. And, moreover, the decisions to use particular quantities of copper and of iron, coal and all the other materials, manufactured products and services for the construction and operation of the generators and the rest of the generating station must be linked by their prices to their respective sources of supply and production.

The benefits of an application of technology are dissipated if those links are broken. Thus if the prices of coal and of copper conductors are fixed arbitrarily without regard to the processes of mining and transporting the coal and of mining, smelting and transporting the copper and turning it into conductors, the design of the generating station will become equally arbitrary and its contribution, if any, to the prosperity of the community will be a matter of luck.

Such a rational pricing system does not arise and persist automatically in the commerce of an industrial society as it grows in size and complexity. Contrary to the apparent belief of some economists, it is not embedded in every so-called 'free-market economy'. Some forces support it but there are others which oppose it and tend to weaken and break those necessary links. We shall come to them in the next chapter.

The broad principle, which was developed in Chapter Three, is that the contribution of an application of technology to prosperity is greatest when its design is based on minimising its resource costs, which are what the community must forego in order to make that application. But in practice designs are based on minimising commercial costs. This is not the occasion to return to the effects of taxes and subsidies, but to address the problems of how to bring the commercial costs, which each customer must pay to obtain the goods and services of a commercial undertaking, as nearly as possible into line with the respective resource costs incurred by the undertaking in producing or providing those goods and services, when taxes and subsidies are not at issue.

That alignment is seldom, if ever, the main objective of any commercial organisation, partly because the prosperity of a commercial organisation does not go hand in hand consistently with the prosperity of the society of which it is a part (the organisation may thrive at the expence of the society or vice versa), but also because of the complexity of the problems in the path to that objective. The objective should be applied to every commercial organisation, but this chapter will be restricted to commercial public services in public ownership. In as much as nearly all the commercial public services in Britain have now been privatised, that may sound like an academic exercise, but private ownership and competition are complicating factors which it is convenient to leave on one side in the first instance. Moreover, commercial public services are still largely in public ownership in some countries and are likely to remain there. The problems of private ownership and competition are tackled in the next chapter.

It would be perverse to insist on a principle and close off possible loopholes of escape from it if it cannot be applied. On the other hand, it would be misleading to gloss over the difficulties. Two purposes of this chapter are to show that the problems are real and large, but not insurmountable. They are rooted in the technology employed by each organisation, which is often not thoroughly understood – or, if understood, not fully taken into account – by those who are responsible for the commercial methods and policy of the organisation. Those roots make the problems different in different sectors of an economy, but one thing public services have in common is substantial investment in tangible fixed assets – to use the accountants' somewhat clumsy but established term. This chapter will of course make use of the theory of annual depreciation of fixed assets developed in Chapter Four and it will assume that the annual accounts are – or can be – Adjusted Historical Cost Accounts, as described in Chapter Five.

We shall concentrate mainly on public electricity supply for the reasons stated in Chapter One (page 10) and assume that the supplier is neither subsidised nor taxed, except as a commercial undertaking like any other. How should such a public electricity supplier frame its tariffs (sometimes known as 'rates') so as to charge its consumers the respective true[2] resource cost of supplying them, within the limits of accuracy of its data?

The benefits of charging the true resource costs and the harm of charging more or less than those resource costs are shared between the consumers, the non-consumers who are excluded or discouraged by the tariffs and the monopoly supplier. And the tariffs affect the decisions of technologists and others who use electricity – or decide not to do so – and the decisions of the supplier's technologists. But of course the supplier's costs are what it pays its own suppliers, contractors and work force. It is therefore a prerequisite for charging its consumers the true resource costs of their electricity supplies that the electricity supplier shall itself be charged the true resource costs by its own suppliers, contractors and work force. And it is a prerequisite for those suppliers etc. charging the electricity supplier the true resource costs of what they supply that they shall in turn be charged the true resource costs by *their* suppliers etc. And so on *ad infinitum* – or perhaps *ad nauseam*.

If we were easily daunted, we might give up at this point in the expectation that investigating all those resource costs would be a wild goose chase, and moreover that some of the geese – such as the cost of aluminium for cables – would be found roosting in the rate of growth of electricity supply systems (see Chapter 2, page 31), which will depend on the tariffs we have not yet designed. However we can circumvent the difficulty by assuming, *ab initio*, that the electricity supplier is charged the true resource costs of its goods and services. That means that initially the contribution of public electricity supply to our prosperity will be less than it might be to the extent that that assumption is wrong, but that is not a sufficient reason for abandoning our investigation. It would only be a sufficient reason if we knew or strongly suspected that modifying the tariff for, say, aluminium production plants, which use large quantities of electricity, would affect the price of aluminium so much as radically to alter the cost of electricity and so upset that modified tariff. Technologists might describe such an outcome as a 'destabilising feedback', but there is no reason to fear anything like that because, although the cost of electricity looms large in the cost of aluminium, the cost of aluminium is only a very small component of the cost of electricity, and it would still be only a small component if aluminium completely superseded copper for the conductors of generators. On the other hand, although the cost of coal looms large in the cost of electricity and coal mines depend on electricity from the public supply, the cost of electricity is only a small component of the cost of coal.

Every investigation has to start somewhere and electricity tariffs are a good place to start this one, partly for the very reason that electricity supply is not so closely double-linked to any other industry as to create any such destabilising feedback. If we are successful, our methods can be adapted and applied in other industries and sectors of the economy, until our initial assumption gradually becomes a true statement. The similarities and differences between public electricity supply and some other public services and their effect on their pricing systems are reviewed at the end of this chapter.

TWO ECONOMIC THEORIES

Average cost pricing

There are numerous extant methods of charging for electricity supplies, but they are all based on or derived from one or other of two economic theories, or on compromises between them. The theories are known as 'average cost pricing' and 'marginal cost pricing' but neither is satisfactory from the point of view of deriving prosperity from technology. Let us trace and comment on the reasoning which leads to average cost pricing, step by step. It goes as follows:

1) The job of a public electricity supplier is to supply electricity to consumers who want it and are willing to pay for it. – Agreed.
2) Electricity is measured in kilowatt-hours. – Yes.
3) The supplier incurs costs in generating and distributing those supplies. – Indeed.
4) And it can forecast with sufficient accuracy: (i) how many kilowatt-hours will be consumed in total next year and (ii) the total of its annual costs next year. – Well yes, with one proviso, a competent supplier should be able to do so, but . . .
5) (interrupting that comment) Very well then, the supplier should divide the estimated total cost next year by the estimated total consumption next year to obtain an average cost per kilowatt-hour, and should charge that average cost as a flat-rate price to all the consumers next year. – Well no, that really will not do!

The objection to average cost pricing is fundamental. It stems from the very reasons why consumers generally find it cheaper – usually much cheaper – to buy their electricity from the public supplier than to generate it themselves, as would be technically feasible in every case. There are four main reasons:

1) By combining the demands of many consumers, the public supplier obtains the advantages of large-scale generation

with plant which is cheaper to acquire and to operate than all except perhaps the very largest consumers would have to employ.

2) To an important extent the consumers use electricity at different times and the public supplier can nearly always obtain more reliable service from its plant than most individual consumers, so the supplier requires less total generating capacity than the consumers would require in aggregate if they generated their own supplies.
3) The public supplier can locate its generating stations economically, where fuel or water power, cooling water and/or suitable land are comparatively cheap, and transmit electricity in bulk to the load centres.
4) The total demand for electricity fluctuates widely from hour to hour, day to day and season to season, so the public supplier can co-ordinate the operation of its generating stations to minimise their total operating cost and it can design its stations to obtain the greatest benefit from such co-ordination.

Those advantages are partly offset by the necessity of constructing and operating electrical networks to interconnect the generating stations, transmit their outputs in bulk to the load centres and distribute them among the consumers, and the consumers are not equally dependent on those networks. Consumers remote from the generating stations are more dependent than those near the generating stations. Moreover in any particular location, small consumers are more dependent than large consumers because they gain more from the advantages of large-scale generation and less total generating capacity by the public supplier[3].

Consequently if every consumer were charged at the same flat rate per kilowatt-hour, the large consumers near the generating stations would find it was cheaper to generate their own supplies and would do so. That would increase the average cost and so the price to the remaining consumers, which would lead to more consumers opting out of the public supply and further price increases until the system collapsed, leaving most of the consumers with no supply or connected to small networks at prices so high that they would use electricity only for lighting at low levels of illumination and a few low-consumption appliances. In effect the history of public electricity supply would be reversed, but at an unpredictable rate. The interrupted proviso, in the comment on the fourth step in the foregoing reasoning leading to average cost pricing, would have been – paradoxically – that the supplier shall not be required to adopt average cost pricing. If it were, it would no longer be able to estimate the total consumption and cost next year, with any pretence of accuracy, in the collapse which would ensue.

There are probably few serious champions of average cost pricing for electricity supplies nowadays, although it is easy to be tempted in that direction when the public supply system is growing and appears to be able to recover its costs from its customers with no difficulty. But there are many influential advocates of it, often on putative moral grounds, for other industries and sectors of the economy, notably transport and postal communications. It casts some light, albeit obliquely, on the economic problems of those sectors to have it spelled out how average cost pricing would wreck the electricity supply industry and impoverish us all in the process. But if not average cost pricing, what then?

Marginal cost pricing

The other theory, marginal cost pricing, is that the price charged should reflect the cost to the producer of an additional unit of output, that is an additional kilowatt-hour, at the margin of production. The classical version of the theory stems from the variations in the fertility of land under primitive agriculture, but when it is applied to applications of modern technology, an important difference between natural and created wealth seems often to be overlooked – or at least not followed up.

If a primitive agricultural community has plenty of land, the classical theory is that that community first cultivates the most fertile land, typically in the valley. As the population grows, more food is needed and so farms are started on the hillsides, where the land is less fertile. On those farms more labour is required to produce any given measure of food, such as a bushel of wheat or a litre of milk, than in the valley; consequently the farmers on those 'marginal' farms require more income to stay in business and must charge higher prices per bushel and per litre than the valley farmers needed to charge when they could produce enough food for the whole community. But the valley farmers can now charge as much as the hill farmers for food and, in an economy which relies on money rather than physical controls to regulate what people will do and buy, valley farming produces a surplus of income over expenditure which is not related to any effort or risk. It accrues to the owners of the more fertile land, whether they cultivate it themselves or rent it to tenant farmers. Such surpluses have of course been sources of social resentment, leading to political conflict, and often to revolution and misery, throughout the history of agriculture.

The application of that classical theory to public electricity supply is unexceptionable in two minor respects, namely that it can include higher prices for remote consumers than for near consumers to pay for transport (which is akin to transmission), and higher prices for small consumers than for large consumers to meet the higher costs of distribution, administration and accounting for small consumers. It can also be elaborated to include seasonal variations in prices. But it overlooks the fact that fertility, which determines how much labour is required on a farm, is an intrinsic property of land under primitive agriculture, whereas the efficiency and other physical attributes of a thermal generating station, which determine how much fuel and stores it consumes and how many operating staff it requires, are designed properties of the station. Within wide limits, as described in Chapter Two (page 39), the designers can increase the efficiency and economy of a generating station at the expense of increasing its capital cost. Hydro-electric stations are in a different category in as much as their attributes are largely determined by their locations, which are limited in number and suitability. Hence they form only a very small part of the total capacity in most industrial countries and can be fitted into a pricing system which is based on thermal stations, as will be explained later in this chapter.

If the farmers in early agricultural societies had been able to design their land in the same way as modern technologists can design thermal generating stations, the history of agriculture, wealth distribution, government and human conflict might have been very different. The fact that modern farmers can make their land substantially more fertile by applying synthetic fertilisers, which affects the shape of modern farming economics, is beside the point because the fertilisers did not arrive until most of that history had been written. It is the relevance of the theory to thermal generating stations which needs to be particularly scrutinised.

THE CO-ORDINATION OF ELECTRICITY GENERATION

To facilitate that scrutiny, we need what technologists, economists and others call a 'model' of an electricity supply system. A model railway gives its owner an idea of how a real railway works and quite small models have been successfully used in the design of real railways. For example, experiments on small-scale models of the overhead conductors and their supporting structures, from which electric trains pick up their power supplies, have shown how the pick-up gadget on the roof of the train (known as a pantograph) can be prevented from parting company with the overhead conductor when the train is running fast. More generally a model is anything, not necessarily a physical thing, which helps one to envisage and calculate how something will work without carrying out a full-scale experiment on it.

For our purpose we need not a physical model but an economic model to tell us how much it will cost to construct and operate an electricity supply system under any of a number of conditions we have in mind. That is a very large order and we shall not attempt to execute it all at once. Fortunately we can simplify it by remembering that the electrical networks, which interconnect the generating stations, transmit their outputs to the load centres and distribute them among the consumers, were described (on page 196) as necessities which offset the four advantages of public electricity supplies. To a casual observer who does not live or work near a generating station they may be the most noticeable (some would say objectionable) feature of an electricity supply system, but economically they are still only of minor importance.

In Britain, in 1988/89 (the last year when electricity supply was publicly owned) generation still accounted for about 80% of the total annual cost, including the cost of the energy lost in the networks, and that figure was not expected to fall much in the near future. So we can begin by leaving the costs of the electrical networks on one side and concentrate on the costs of generation.

A suitable economic model was introduced in Appendix 2 of Chapter Four (page 146) for the purpose of estimating the economic lives of groups of fixed assets providing a common service, such as generating stations. It listed five physical measures of the annual utilisation of a generator, but a single measure, namely the annual output (measured in kilowatt-hours) is sufficient to understand the principles involved. The model then has two components, called the 'running cost' and the 'standing cost'. It can be applied to any generator whose costs are kept separately in the accounts, but it is more convenient to apply it to whole generating stations whenever (as is usually the case) the generating units (generators, turbines and boilers) are the same size and design, and so have the same, or almost the same, costs.

In the simple model, the running cost of any station, which is approximately proportional to its annual output, comprises the costs of its fuel and stores, together with the cost of its shift operators. The standing cost, which is independent of its output, comprises the remaining staff costs and the interest on and depreciation of its capital cost. Thus the running cost is equal to the operating cost, as normally defined, minus those remaining staff costs, which are part of the standing cost. Let us call the running cost per kilowatt-hour and the standing cost per kilowatt of output capacity (that is the output which cannot be exceeded for any length of time) the 'cost parameters' of a generating station. They can be calculated from the output record and cost accounts of a station in operation or estimated from the designed properties of a projected station.

With those simplifications, there is a straightforward technique for operating the generators and stations to the best advantage as the total demand fluctuates hourly, daily and seasonally. They are first listed in order of increasing running cost per kilowatt-hour, which is called the 'merit order', that is to say the station with the lowest cost per kilowatt-hour is put at the head of the list and the station with the highest cost per kilowatt-hour is put at its foot. Then, as the total demand increases, the stations and the generators in them are started up and loaded up in that merit order to meet that demand with as few generators as possible in operation at any one time, after making adequate allowances for accidental breakdowns in the generating stations and/or parts of the networks and for unexpected, rapid increases in the demand. And as the demand decreases, the generators and stations are of course off-loaded and shut down in the reverse order.

When the model has more cost parameters and account is taken of the costs of, energy losses in and vulnerability of the networks, the technique becomes correspondingly more elaborate. It was employed to great advantage when the British grid was constructed in the 1920s and 1930s, even though the several parts of the grid were seldom connected together. During and ever since the Second World War, the whole grid has been operated as a single, continuously connected entity, which has given correspondingly greater scope for economies, and the advent of computers has enabled refinements to be made which were previously not possible[4]. One such refinement was to take account of the changes in the prices of fuel to the stations, which varied from month to month, so altering the merit order of the stations during the year.

Bearing in mind that in 1988/89 the cost of fuel alone was more than £5 billion (60% of the total cost), it is obviously well worthwhile to employ the best operators and computers and the cleverest available programmers to minimise the total running cost of the system – or at least it was until the system was split up and privatised, but we shall come to the effects of privatisation in the next chapter. The only limitation is the accuracy of the data, i.e. of the cost accounts of the stations and networks and of the estimated costs of projected stations and of projected extensions of the networks. However our concern is to understand the principles of the technique and that it is capable of elaboration and refinement. We need not delve into all its complexities.

When the theory of marginal cost pricing has been applied to electricity tariffs, and in other industries, a distinction has commonly been drawn between the 'short-run marginal cost', or 'SMC', and the 'long-run marginal cost' or 'LMC'. In the present context, the SMC is nearly enough the same as the running cost parameter of the station of lowest merit which is in operation at any one time. It varies from season to season, from day to day and from minute to minute as stations are started up, loaded up, unloaded and shut down to meet the varying total demand. In the analogy with primitive agriculture, the SMC of electricity corresponds to the cost of producing food from the least fertile farm, which is brought into production when the population grows to the point where it can no longer be fed from the more fertile farms, and is taken out of production if and when the population declines.

The demand for food grows and sometimes declines from year to year, whereas the demand for electricity varies from minute to minute, but the analogy holds good in as much as the SMC at any moment in an electricity supply system is inevitably greater than the running costs per kilowatt-hour of the stations with lower running-cost parameters in operation at that time. So, if the price is equated with the SMC, the income accruing to the stations nearer to the head of the list produces a surplus over their running cost, which may appear to be analogous to the surplus enjoyed by the valley farmers – or their landlords. But that is before any account is taken of the capital costs of the stations, which are resource costs having no true parallel in primitive farming.

The LMC is intended to include the capital costs of the generating stations, which also vary from station to station. But attempts to calculate the LMC per kilowatt-hour of output from the stations in operation at any one time have been frustrated by the lack of any agreed, rational method of apportioning the capital cost of a generating station between the years of its economic life, that is to say by the lack of any adequate theory of annual depreciation. That lack has led to a confused situation, which was paraphrased as follows at an international conference on electricity tariffs in 1987:

> One aspect ... is the theory of marginal cost pricing. ... Economists try to maximise welfare and take account of future costs, accountants record and audit historical costs incurred. These different professional practices have to be reconciled by management when, having formulated tariff prices in accordance with economic theory (where possible), the projected financial out turn has to be met. No theory will do this: it is simply a matter of sensible massaging of the tariff elements until an acceptable out turn is projected.[5]

'Massage' in that context is a graphic synonym for 'compromise', which is of course always to be preferred to violent revolution. And as long as electricity was cheap and becoming cheaper and its generation did not appear to harm the environment, compromises of that kind did not appear to threaten our prosperity, but that euphoria is fast evaporating. Its disappearance warrants a fresh search for an adequate theory of electricity tariffs, for which we are now well equipped with a theory of annual depreciation. To prosper from technology we must avoid both the cauldrons of social resentment and the debilitations of the financial massage parlours.

In Chapter Two, bled-steam boiler feedwater heating was described as a zenith of technical complexity (page 32). We are now on the edge of a forest of economic complexity, which becomes thicker and darker as one goes into it. It is commonly regarded as impenetrable, so it should occasion no surprise that some intellectual effort will be required to get through, even without the mathematical formulae. They are set out in the two appendices to this chapter, but the objective is to show that the difficulties can be overcome, not to present a complete package of electricity supply tariffs in detail, which would be tedious.

The economic cost of generation

We must return to the question with which we began: 'How can a monopoly public electricity supplier frame its tariffs so as to charge its consumers the respective true resource costs of supplying them, within the limits of accuracy of the data?', with the addendum now: 'assuming that the supplier is itself charged the true resource costs by its own suppliers, contractors and staff.'

The annual capital charge (interest plus depreciation) constitutes the greater part of the standing cost of a station and the first branch of the theory of depreciation developed in Chapter Four (page 123) enables the charge to be calculated approximately from the capital cost of the station and its estimated economic life. But the wedge formula in that first branch takes no account of the changes in the station's annual output from year to year, although they are important consequences of co-ordinated generation and so must figure in the electricity tariffs. We need the second branch, which does take account of those changes.

Appendix 2 of that chapter (page 152) explains how the 'economic cost' of every asset in a group which is used to produce or provide a common good or service – in this case a group of generating stations – can be calculated. It represents the total annual cost of producing the output of that station by the most economic available means, with the stipulation that the calculation must provide *in toto* for the depreciation of all the stations in that year.

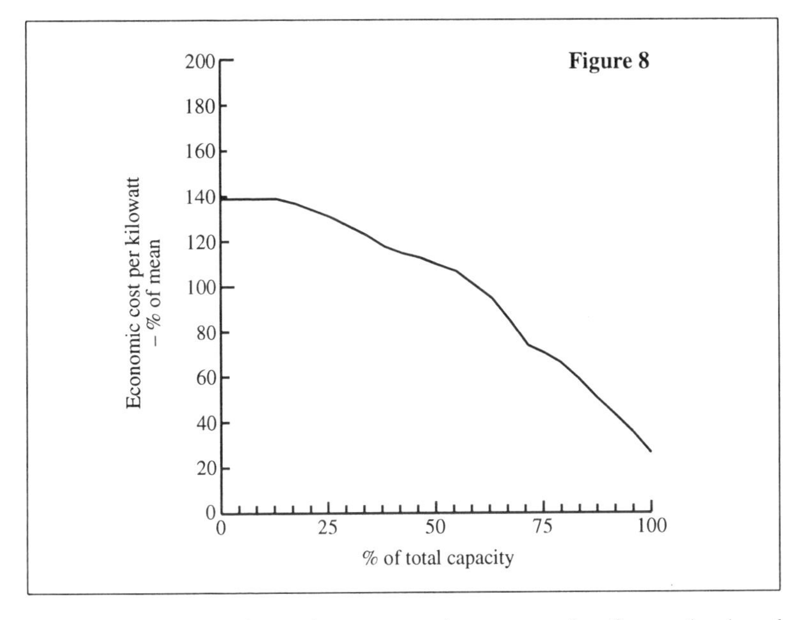

Figure 8 shows how the economic cost varies from the head to the foot of the merit-order list in a typical year. The important feature is not the absolute magnitude of the economic cost at any point in the list, nor the total capacity of all the stations, both of which have been eliminated by the choices of the variables, but the shape of the graph. From the way the quantities have been calculated, that shape can change only slowly as the stations are constructed and commissioned, pass through their long service lives and are eventually scrapped. It would not be affected by any recalculations of the economic lives of the generating stations because such calculations would simply increase or reduce the total of the capital charges without affecting the differences between the economic costs of the individual stations. The shape is therefore characteristic of the supply system.

The economic cost of any individual station is not generally the same as the cost which could be obtained directly from its annual accounts, although the total economic cost is equal to the total cost of all the stations. It remains to consider the causes of the differences and to show that the relevant costs in framing electricity tariffs are the economic costs. Those points will be tackled, *inter alia*, in the next section. Meanwhile there is a loose end to be picked up.

Hydro-electric stations

Our economic model was devised to represent thermal generating stations, which include coal-fuelled, oil-fuelled, gas-fuelled and nuclear stations but not hydro-electric stations – commonly known as 'hydro' stations. They have high capital costs and therefore high standing costs, and correspondingly low running costs because the water is free once the stations are built. Most supply systems have a few hydro stations and in a few systems they are dominant.

Hydro stations are not operated in the same way as thermal stations, according to their running-cost parameters; indeed they cannot be because their outputs are physically limited by the capacities of their water reservoirs. Generally a hydro station has a dam, which creates its reservoir, and the sizes of the dam and the reservoir are limited by the local topography and the high costs of construction, but in comparison the turbines and generators are cheap. Consequently, in applying the cost method, the designers soon find themselves balancing the extra cost of larger turbines and generators against the advantage of being able to generate more of the limited output around the times of peak demand, when it will displace the output of some correspondingly higher-running-cost thermal stations. And so the hydro stations finish up with turbines and generators which can empty their reservoirs in a few days, or even a few hours.

The output also depends on the rate at which the reservoir fills up again, which depends in turn on the rainfall – a high rate in the rainy season (which may or may not coincide with the season of high total demand for electricity) and a low or zero rate in the dry season. Thus operating hydro stations involves some fairly complicated juggling. How should they be included in the calculations of economic cost?

Going back to the thermal stations for a moment, the merit-order list had them in order of increasing running-cost parameters, but the depreciation calculations put them in order of decreasing annual output per kilowatt of output capacity. Along most of the two lists, the orders are the same, but there are bound to be discrepancies. When it comes to the point, the order of decreasing annual output takes precedence in calculating the economic costs, which means that, in a few cases, the running-cost parameter of the second station in a pair of consecutive stations is lower than that of the first. Consequently, from the equations, the standing-cost parameter of the second station comes out higher than that of the first station in such a pair.

That minor change of emphasis has the incidental advantage of enabling the outputs of the hydro stations to be included in the list with no difficulty or ambiguity. They are slotted in wherever their annual outputs per kilowatt of output capacity dictate. Now when a hydro station follows a thermal station, which of course has a much higher running-cost parameter, the comparisons of that pair will give the hydro station a correspondingly higher standing-cost parameter, which neatly matches its higher capital cost.

In this context, 'hydro stations' include pumped-storage stations. In a pumped-storage station, a second reservoir is constructed to collect the water which comes out of the turbines, and the turbines are designed and constructed so that they can be run backwards as pumps, using their generators as electric motors. Then whenever the upper reservoir is at a low water level and electricity can be generated elsewhere at a low running cost, usually at night and often in a nearby nuclear station, water is pumped from the lower reservoir back into the upper one, thus increasing the quantity of electricity which can be generated at or near the peaks of the next few hours or days. In our calculations, a pumped-storage station is treated as a hydro station whenever it is generating and as a consumer whenever it is pumping, but its annual cost as a generating station must include whatever it is charged for electricity when it is pumping.

One case has not been covered. If the hydro stations are dominant, the whole business of co-ordinated generation and of calculating the economic cost has to be worked out afresh. But that sort of loose end is not important in considering the principles which underlie – or should underlie – electricity tariffs, because very few supply systems are dominated by hydro stations. A set of calculations which are applicable to the great majority of supply systems is not invalidated just because a few systems might require something different.

It may perhaps be pointed out that nowadays the locations and designs of thermal generating stations no longer depend solely on economic considerations – that both are increasingly affected by land-use planning requirements and other physical restrictions for the protection of the environment. The consequences of such restrictions are considered in Chapter Nine.

THE 'DEARNESS' OF ELECTRICITY

A set of tariffs which charge the consumers the true costs of supplying them must comply with three conditions. They may seem rather curious and they are certainly hypothetical, but in combination they provide a sufficient framework for the tariffs.

The first condition is derived from the case of two hypothetical consumers, side by side in the network, whose demands are equal, instant by instant, throughout the year. The supplier's costs of meeting those demands will be the same. The uses to which the consumers put their supplies are irrelevant. One might bake bread for the needy poor while the other cultivates orchids for the idle rich; the costs of supplying them are still the same. *Ergo* they must be charged the same amount. That is the first condition. It amounts to saying that the only factors which can be allowed to affect the tariff are the consumers' positions in the network and the magnitudes of their demands at every instant. Readers who feel that this statement is immoral or that it omits a vital moral issue are invited to turn to the next main section of this chapter and to Chapters Eight and Ten, where such issues are considered.

The supplier's costs are the same

Now let us suppose that the second consumer is replaced by a number of smaller consumers, the total of whose demands happens to coincide at every instant with the demand of the first consumer, although the individual demands of the smaller consumers are quite different from one another. Of course the costs of connecting the smaller consumers and of meter reading, billing, extending credit and so on will be larger in aggregate than for the single large consumer, but those considerations are red herrings which can be removed from the trail by stipulating that those costs will be charged separately. Then the total cost, excluding those separate costs, of supplying all the smaller consumers will be the same as the cost of supplying the single large consumer. The tariff must therefore make the sum of the smaller consumers' payments the same as what the large consumer pays, irrespective of how their demands vary individually. That is the second condition.

To comply with those conditions, it must be possible to calculate the cost of any supply from a quantity which varies from instant to instant and from point to point in the network, but is the same for every consumer at any one point of supply[6]. To denote such a quantity we need a term which conveys the idea of cost and is measured – as a price is measured – in pence (or cents or whatever) a kilowatt-hour, but does not imply that the consumers will necessarily pay that price, although it will be the starting point for calculating what they will pay. A suitable non-committal term, which also has the advantage of no prior commercial associations, is the 'dearness' of electricity.

If the dearness at a particular moment and point of supply is, say, 5.4 pence a kilowatt-hour, then the resource cost of supplying, say, 58.7 kilowatts at that moment and point will run at the rate of 58.7 × 5.4 pence = £3.17 an hour. In the ordinary way the demand will vary from moment to moment and so will the dearness, independently. If we know how they both vary, we can multiply them together and so calculate the cost of the supply over any period, just as we can calculate how far a car will travel if we know how its speed varies from moment to moment[7].

The third condition is derived from the case of a hypothetical large consumer whose demand happens to be the same at every instant as the output of an immediately adjacent generating station, operating under co-ordinated generation. The cost of that supply is the same as the economic cost of that station, as calculated in Chapter 4 (page 152). It has to be the economic cost, rather than the cost in the annual accounts for the station because the stations in operation at any one time were necessarily designed up to 25 years earlier, on forecasts then of the total demand and its daily and seasonal variations. Such forecasts are inherently uncertain and the errors are compounded by imperfections in the designs – sometimes slight, sometimes grave. Consequently the annual cost in the accounts of any particular station may be as low as, nearly as low as or much higher than the minimum which might have been achieved with exceptionally fortunate forecasts and the best design of that station. The difference is a matter of luck; it cannot be foreseen.

Now consumers with demands which match, even approximately, the outputs of generating stations operating under co-ordinated generation are particularly well placed to generate their own electricity and will tend to do so, to the disadvantage of the community, if they have to pay for the cost of the public supplier's worst efforts. The consumers as a whole have to stump up for the impossibility of exact forecasting and for the public supplier's errors, and they benefit from the supplier's best efforts, but the unforeseeable costs should be distributed among all the consumers, so as to minimise the inducement to any consumer to opt out of taking a public supply merely because of an unfortunate coincidence of forecasting and design errors.

A model of the dearness

So how can we calculate how the dearness must vary to conform to those three conditions? We need another model and it can be built in stages with the help of a mathematical technique, which is explained in Appendix 1, to show how we are getting on at each stage. The technique works by telling us how well the model fits the data at whatever stage we have reached. There would be no point in striving for a perfect fit because our data arc far from perfect.

The simplest conceivable model would make the dearness constant throughout the year. Its value would then be obtained by simply dividing the total cost of generation (80% of the total cost of supply) by the total number of kilowatt-hours supplied. In 1988/9, that would have worked out at 4.1 pence a kilowatt-hour.

In effect that is a model for average cost pricing. If the constant dearness were multiplied by the annual output of any station, the product would represent the resource cost of supplying a consumer whose consumption exactly matched that output throughout the year – according to the model. Let us call that product the 'dearness worth' of that output. Ideally the dearness worth should equal the economic cost of the station, so a comparison of the dearness worths of all the stations with their respective economic costs will show how well the model represents the data.

Figure 9 presents that comparison visually. It shows that, at the base-load end of the range (at the left hand side of the figure), the dearness worth is substantially larger than the economic cost and at the peak-load end it is very much smaller. Hence, even without bringing the cost of and losses in distribution into the calculation, a set of tariffs based on that model would induce at least some large consumers at the base-load end to generate their own electricity, while the subsidies at the other end would tend to inflate many small consumers' demands, leading to the eventual collapse of the system as already described (page 196).

That model fails because it takes no account of the designers' scope for reducing the economic cost of generation by increasing the capital cost of – hence the capital charges on – the stations, especially at the base-load end of the range. To include that scope, let us try making the dearness rise and fall in direct proportion to the total demand. Its magnitude then works out at 3.6 pence a kilowatt-hour (at 1998/9 price levels),

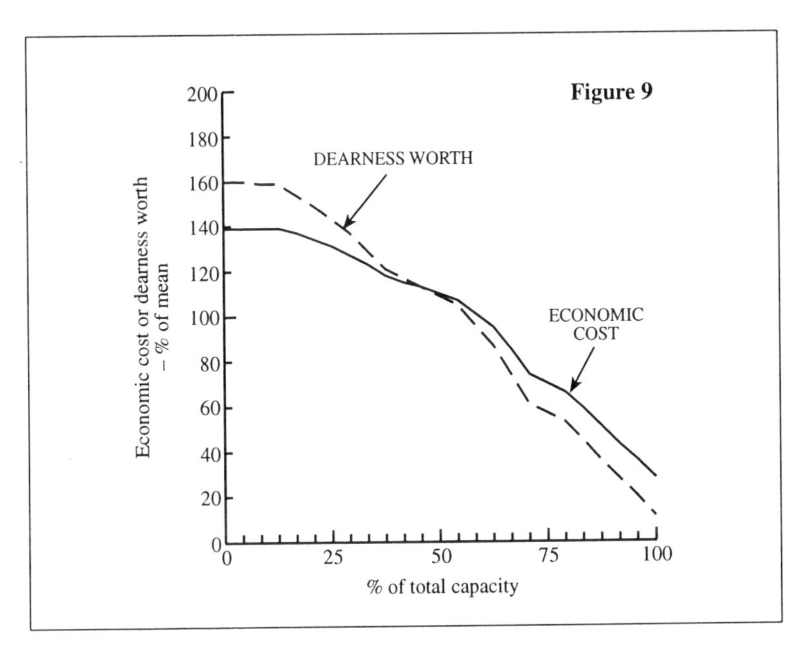

Figure 9
200
180
160
140
120
100
80
60
40
20
0
Economic cost or dearness worth
– % of mean
DEARNESS WORTH
ECONOMIC
COST
0
25
50
75
100
% of total capacity

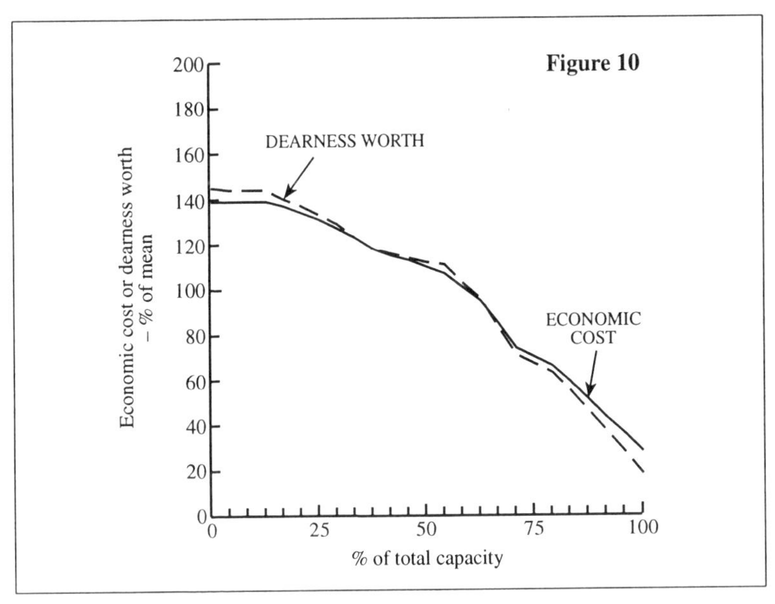

Figure 10
200
180
160
140
120
100
80
60
40
20
0
Economic cost or dearness worth
– % of mean
DEARNESS WORTH
ECONOMIC
COST
0
25
50
75
100
% of total capacity

multiplied by the ratio of the total demand at any instant to the average total demand during the year. That ratio varies from about 40% to 200%, so the dearness varies from about 1.5 pence a kilowatt-hour when the total demand is at the minimum level (in the middle of the night in midsummer) to 7.1 pence a kilowatt-hour at the time of the highest peak in mid-winter.

Figure 10 compares the dearness worths of all the stations on that basis with their economic costs, in the same way as Figure 9. It shows that, in spite of its simplicity, this second model equates the dearness of every station to its economic cost to a remarkably close approximation. In other words the model seems to comply well enough with our three conditions and so to provide the necessary framework for a set of tariffs which will charge the consumers the true generation costs of supplying them – but there is a pitfall ahead.

We could construct a more elaborate model, with two or three components as explained in Appendix 1, which would fit the dearness worths to the economic costs even more precisely, but the errors in the data would probably swamp the small differences. Let us therefore continue with this model, while retaining the option of elaborating it as and when better data become available and we have avoided the pitfall.

Network costs and losses

At some stage, the costs of and losses in the networks between the generating stations and the consumers must be incorporated into the model. As previously mentioned, the networks have three functions: (1) to interconnect the generating stations so that their outputs can be co-ordinated, (2) to transmit their outputs in bulk from the economical sites of the stations to the load centres and (3) to distribute the outputs to the consumers. The total annual cost of all three functions together amounts to about 20% or so of the whole annual cost of the public supply and so represents an increase of 25% (i.e. 20/(100–20) %) on the cost of generation. The losses amount to about 5% of the generated output.

The costs and losses of interconnection and bulk transmission effectively increase the dearness to all the consumers and, as a first approximation, we can allocate half the network costs and losses to those two functions by increasing the single parameter in our model by 15%, that is from 3.6 to 4.1 pence a kilowatt-hour. The other half will create differences between the different parts of the networks, but that analysis will not be pursued in this book.

Forecasting the dearness

So far the calculations have been retrospective. At the end of a year, the total output, the economic costs of the stations and the networks and the parameter of the dearness model (4.1 pence a kilowatt-hour) can be calculated from the records of the total demand, the outputs of the stations and their annual accounts and the costs of and losses in the networks. But of course the consumers need to be able to work out in advance how much they will be charged, so that they can regulate their demands to their own best advantage.

The essential requirements are that parameter, which will change only slowly in real terms from year to year, and the best possible forecast of the total demand. Now the total demand is affected by – indeed largely regulated by – the alternation of night and day, the passing of the seasons, the weekly cycle of working and leisure and the occurrence of public holidays and quasi-public holidays[8]. In other words the total demand rises and falls largely in cycles or waves, of which the amplitudes and the times of the crests can be measured from past records and used to make the forecasts, as described in Appendix 2.

Figure 11 shows how the forecast total demand varied during a winter day, compared with the actual total demand on such a day. The forecast happened to be lower than the actual demand on that day but it would of course be higher on other days. Also the small fluctuations from hour to hour are not in the forecast because they are essentially random and so unpredictable. (It is important not to forget that the purpose

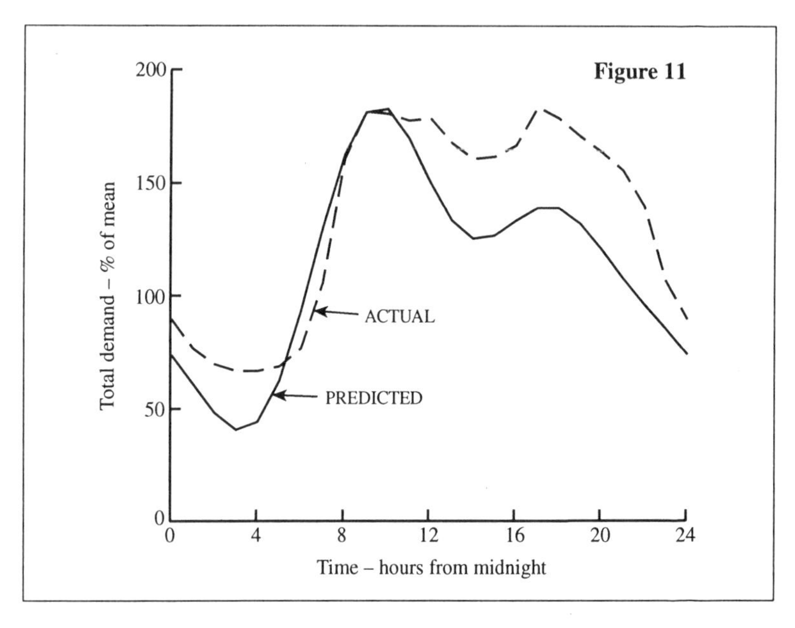

of these forecasts is to provide a framework for the tariffs, not to plan how the demand can best be met.)

Corresponding forecasts could be made for all the other days of the year. However the dearness, calculated from such forecasts, does not represent the price which would maximise the community's prosperity from public electricity supply. That is the pitfall anticipated above.

The reason is, of course, that the total demand is affected by the tariffs, particularly by large changes to the tariffs. Generally an increase in the price of electricity causes a reduction in the demands of affected consumers and a reduction in the price increases their demands, but the scales of those effects and the speeds of the consumers' reactions are very different for different kinds of consumption. For some purposes, such as lighting, electricity is cheap and practically irreplaceable, so the price has little effect on the consumption in the short term, although it stimulates research into more efficient lighting in the long term. But for other purposes, such as the heating of buildings, the advantages and disadvantages of electricity compared with other forms of energy are smaller, so the consumption is sensitive to the price within a much shorter time-scale. Moreover, some demands can be easily adjusted to take advantage of changes in the times when the price is low, and such retiming affects the waves or cycles on which the forecasts are based. So what is the point of forecasting the dearness?

The dearness and the tariffs

The dearness represents the varying price at which the commercial charge to any consumer, calculated from that price, would be the true resource cost of supplying that consumer if it did not affect the total demand at that time. Hence the differences between those resource costs and the commercial charges, as calculated from the tariffs, provide pointers to the scope for adjusting the tariffs – to the advantage of some consumers and inevitably the disadvantage of others but to the net advantage of the consumers and potential consumers in general.

If the commercial charge to any consumer or class of consumers, as calculated from the relevant tariffs, were considerably greater than the resource cost of supplying them, as calculated from the dearness, that would indicate that their tariff could be adjusted to reduce their charges, so reducing the difference between the commercial charge and the resource cost. They would normally react (after some delay) by increasing their demands, so increasing the resource costs of supplying them and further reducing that difference. And vice versa for any consumer or class of consumers whose commercial charges were less than the resource costs. By such adjustments, cautiously applied to avoid an overshoot, the commercial charge to every consumer could be brought gradually into line with their respective resource cost.

Thus the dearness provides a reference, against which the tariffs for the various classes of consumers – domestic, commercial, light industrial, heavy industrial, agricultural, railways, etc. – could be initially judged. For domestic supplies on the normal two-part tariff (quarterly charge plus so much a kilowatt-hour), for example, meters to record the variations in their demands as well as their annual or quarterly consumptions could be installed (at the supplier's expense of course) in the premises of a sample of consumers, and their charges under the tariff compared with the resource costs calculated from those meters and the dearness. The comparisons might show that parts of the tariff should be increased or reduced, together or separately, or that a different form of tariff and/or a different way of classifying the consumers would be better.

It is impossible to predict in advance how much and how rapidly the consumers would respond to such adjustments, but that sort of difficulty is familiar to any organisation which sets about changing its prices and/or methods of charging. It is not peculiar to public electricity supply. Such considerations limit the pace at which tariffs can be changed without upsetting the balance between the supplier and the consumers, on which the design and operation of the whole system depend. But they are not a reason for delay. On the contrary, the slower the pace at which the consumers are expected to react to the changes, the more important it is to begin to make those changes. As and when they do react, the tariffs may need to be adjusted, either to encourage stronger reactions or possibly to prevent an overshoot which threatens to reverse the differences between the resource costs and commercial charges to some consumers. And at that time, but not before, it may be worthwhile to elaborate the model, as already mentioned, and the forecasts, as described in Appendix 2, and to make a more accurate allocation of the cost of and losses in the networks.

So far we have tacitly assumed that only one tariff will be offered to any consumer. However it was stated in Chapter Three (page 61) that it is sometimes cheaper in the long run to design a process from the start (at some additional capital cost) so that it can operate whenever the commodity it requires is cheapest in the short run. In as much as all forecasts are uncertain, there must often be a case for offering price reductions with the stipulation that the supply may be reduced or cut off when the total demand approaches its peaks, as an alternative to a maintained supply.

Recent technical progress in metering and telecommunications enables such offers to be refined to the point where the consumption of some large installations can be regulated from minute to minute as the total demand rises and falls[9]. Provided the consumers have a genuine choice (an important proviso), tariffs of that kind can reduce the costs of both the consumers and the supplier. The dearness is well suited to framing such tariffs.

DOUBTS AND CRITICISMS

Some sceptical readers, who come to this chapter with the strong, prior belief that electricity tariffs are necessarily arbitrary, may suspect that the data and calculations in the appendices – particularly Appendix 1 – have been massaged to produce the results which the author intended from the outset. The short reply is that there has been no massaging; the results stem from and rely on the theory of depreciation developed in Chapter Four, particularly the second branch of that theory, as explained in Appendix 2 of that chapter (page 145).

When that theory was first published, it was rejected by some leading accountants, economists and others who came to it in the strong, prior belief that (as one of them put it): 'total depreciation is factual but how you spread it between the years must be arbitrary.'[10] But that was like insisting that it must be impossible to predict the tide more than a few weeks ahead – as indeed it was until the astronomical theory of the tides was formulated and applied.

Similarly with electricity tariffs. They are necessarily arbitrary only until they can be based on rational calculations and allocations of the costs (particularly the annual depreciation) of generation – and ultimately of transmission and distribution. A problem is not impossible of solution simply because it is difficult and prior attempts to solve it have failed.

A second set of critics, attacking on the opposite flank, may contend that the theories, or the author's understanding of them, are at fault or that the data are inferior, and that truer or better understood theories, perhaps with better data, would lead to a different set of results. If their arguments are convincing, they will not be repulsed because any claim that the results could be bettered implies that the main purpose of the previous section has been achieved. That purpose was to show that – in a public electricity supply system at least – the commercial costs, which the customers must pay to obtain the goods and services of a commercial undertaking, can be brought gradually into line with the respective resource costs incurred by the undertaking in producing or providing those goods and services. The method of doing so has been constructed with due care and attention, but it is not put forward as the last word on the subject.

A third set of critics may say that it does not matter much whether commercial costs are brought into line with resource costs because many consumers' decisions are insensitive to the cost of the product or service. They may cite instances in which supplies of electricity are taken with scant regard for the cost, and aver that the supplier's policy should be to 'charge what the load will bear', in other words to charge consumers who are insensitive to the price more than those who are sensitive.

One objection to that policy is that it is unfair – contrary to natural justice – but that is not the only objection. The additional technological objection is that it depends on the supplier having far more information about the consumers' and potential consumers' businesses – their habits and preferences, the multifarious uses which can be made of electricity and the details of all the factors affecting their decisions – than it (the supplier) can hope to acquire. One has only to understand that the consumers' decisions often depend on the availability and costs of numerous mixtures of numerous other resources to realise that judging what a load will bear can hardly ever be better than a guess. And poor guesses affecting many people impoverish the whole community.

Even when consumers take quite large quantities of electricity without any apparent regard for the cost, they have often done the calculations on a previous occasion or they are following in the footsteps of other consumers who they believe have done them. It is both natural and sensible to try to avoid long, expensive studies and to rely on one's general experience and instincts and one's awareness of what others are doing in similar circumstances. Consequently, while a policy of charging such consumers substantially more than the resource costs may not have much effect on their consumption in the short run, it will almost certainly reduce the prosperity of both the consumers and the supplier in the longer term, when the consumers – or some of them – come round to checking their decisions.

'But does not basing the tariffs on the total demand constitute a policy of charging what the load will bear?', such critics may ask. 'Surely you are proposing to bump up the price whenever you judge the consumers most want their electricity. That amounts to taking advantage of their known habits and preferences to charge them more than you would otherwise charge them – in fact just as much more as you judge their loads will bear.' The answer is that the higher prices are calculated from and justified by the higher costs of supplying the consumers at those times, not from or by the habits and preferences themselves.

And that is not just a form of words for a debate. It is a real and important distinction in as much as the consumers are free to adjust the times of their demands to reduce their costs. Some will be unable to do so but not all. If the consumers who can adjust their timings do so, they will benefit not only themselves but also the consumers who cannot. The latter consumers will still pay more for their electricity than the former, but not as much as they would have to pay if the tariffs did not induce the former consumers to make those adjustments. Tariffs which are framed to encourage consumers to adjust their demands to their own advantage are different in principle – not just in words – from tariffs which try to take advantage of consumers who are unable to do so.

A fourth band of critics will argue that it is immoral to insist on charging the full cost to every consumer, regardless of their personal circumstances. They are not very strong in public electricity supply, where there has been a tradition of trying to charge the true cost to every consumer, if only one knew how to do so, and where millions of consumers probably believe that their suppliers do know how to do so and not very many feel themselves to be seriously aggrieved. But in other sectors of the economy, notably transport, there is no such tradition and the very few people who try to argue that we should all be more prosperous if travellers and freight transporters were charged the true costs of the services they receive, are regarded with scorn, as advocates of the impossible, or condemned, as advocates of the immoral.

The belief that people should pay what they can afford or say they can afford, or what they can reasonably be expected to afford, for goods and services, rather than what the goods and services cost, stems from a deep mistrust of money as the means of regulating what people will buy and do – or what they should buy and do. Both the mistrust and the belief are sometimes instinctive – even blind – rather than reasoned, but they are also part of a political creed. We shall return to that creed in Chapter Eight.

Last, but not least, are the critics who will opine that the whole discussion of monopoly electricity supply tariffs has been pointless because the Electricity Act, 1989 created three large generating companies in England and Wales and put them in competition with one another and with generating suppliers in Scotland and France. The replies to them are: (1) it will be at least 20 more years before anyone without an axe to grind can judge whether that Act has increased or reduced our prosperity in Britain; (2) generation is a monopoly in numerous other countries and is likely to remain so; and (3) a few generating suppliers with widely scattered stations do not make a genuinely competitive market; it is more accurately described as an oligopoly. That last point will be pursued in the next chapter.

OTHER COMMERCIAL PUBLIC SERVICES

The prominent commercial public services in most industrial economies are the so-called 'public utilities' – water, gas and electricity supply and the telephone and postal services. In Britain the first four have been privatised (and to some extent broken up) in the last decade and the postal service may not remain in the public sector indefinitely, but the utilities in many other countries are publicly owned.

In any case it is appropriate to give some thought in this chapter to what systems of pricing would effectively charge the utilities' customers the respective true resource costs of supplying them, before going on to consider in the next chapter how utilities and other commercial organisations can be induced to make their best contributions to prosperity when they are in the private sector. It is beyond the scope of this book to develop such pricing systems for the other public services, or any of them, in any detail but there are technical similarities and differences between the utilities which will create corresponding similarities and differences between their pricing systems and so reduce the labour of developing them from first principles.

Firstly all the utilities are capital intensive, so the theory of depreciation developed in Chapter Four will be strongly relevant in every case, and of course Adjusted Historical Cost Accounting developed in Chapter Five. Water and gas supplies have distribution networks and their sources of supply can be interconnected (although it is not generally economical to interconnect water supplies as extensively as electricity generating stations), so the concepts of 'economic cost' (page 204) and 'dearness' (page 208) and the conditions which determine how the dearness will vary from time to time and place to place apply to water and gas supplies. However water and gas can be stored, whereas electricity cannot, so that the dearness of water and gas will fluctuate much less than the dearness of electricity and the calculations will be correspondingly simpler.

There are other, less obvious differences which, if not understood, can cause anomalies and confusion. Thus comparing electricity supply and telecommunications, it might appear at first sight that all the subscribers on a telephone exchange should be charged the same annual rental per line to the exchange because every line 'ties up' the same amount of equipment (apart from small differences in the distances from the exchange), irrespective of how much or little it is used. But, whereas small electricity consumers are more dependent on the network and gain more advantage from the economies of modern large generating stations than large consumers (see page 196), telephone subscribers who make only a few local calls gain less advantage from modern telephone technology than those who make many calls on the national and international networks.

Consequently such a pricing policy might lead to poor people having to give up the telephone. If they did, the value of the network to the remaining larger subscribers would diminish because they would no longer be able to call the people who had given up their phones. Pursuing that line of argument led the Office of Telecommunications (Oftel) to feel at one stage that people with low incomes should be given a special tariff so that they could afford to keep their phones, but that seemed to be at variance with the principle of charging every subscriber the true cost of his or her service[11].

The way out of that apparent impasse was to realise that part of the capital cost of a sophisticated modern exchange (including research and development costs) is justified by the lower cost per call when it is heavily used. That being so, those very small subscribers might be better served for their few local outgoing calls by simple manual or low-cost automatic exchanges, even though their calls would be comparatively expensive per call. But it would be expensive for the other users to make separate connections to those simple exchanges. In that case the interests of all the users, not just the small users, would be best served by connecting them all to the same type of modern exchange, but then charging the small users (irrespective of their incomes) a much smaller fixed rental and a correspondingly higher charge per outgoing call, or per unit of calling time, for an initial block of calls, thereafter reducing the rate to the normal rate for additional calls. Under such a tariff, subscribers with low usage of the telephone would probably not give it up. They would remain connected to the exchanges and so available to be called by other users, but the poor among them would be spared the indignity of having to declare their incomes in order to get what many of them would regard as a humiliating charity handout.

If that tariff, when worked out in detail, could be expected to retain the very small subscribers, it would probably not be worthwhile to try and calculate their value to the other subscribers, arising from the small subscribers' availability to receive calls. But if not, a zero rental might be justifiable, not because the small subscribers were poor but because they would be retained; in fact the retention of a not-so-poor, very small subscriber would probably be more valuable to the other subscribers than the retention of a very poor one[12].

Turning to water supply, in Britain fresh water is at last becoming – or being recognised as – a valuable commodity. Schemes are beginning to proliferate for householders to use less water to flush their toilets, to flush them and/or water gardens with waste water from their baths, for factories and large office blocks to re-use or sell their effluents for non-drinking purposes, for expensive repairing of leaks in distribution pipes, for desalination of sea water (already practised in some hot, dry countries) and so on. But it will be impossible to assess the relative merits of such schemes unless and until a resource-cost value can be placed on the water. And that can only come about if the consumers' consumptions are metered and they are then charged according to their true resource costs of collection, storage and distribution. If that does not happen, it is safe to predict that uneconomic schemes will be introduced and economic ones overlooked and we shall all become poorer than we need be as a result.

Evidently the science and art of framing tariffs for public services are as yet in their infancy, largely because the relevance of their technologies has not been fully appreciated. Looking to the future, it is a matter of common experience that publication is the essential lubricant of progress in both science and art; secrecy is like sand in the bearings. So it is important that the details of all the costs as well as the tariffs should be published, first because the data are needed to frame the tariffs and second to ensure that the services then adopt those tariffs.

There are some valid arguments for permitting genuinely competing organisations to keep some of their commercial information confidential, though the arguments are seldom presented explicitly and are weaker than is often assumed, but those arguments do not apply to monopolies.

REFERENCES AND FOOTNOTES

1) Not forgetting that, in this book, as mentioned in Chapter Two (note 3), a generator is a machine which generates electricity, not a commercial company which produces it for sale.
2) In the sense of 'unmistaken', with no overtones of faithfulness or other moral attribute.
3) In Britain in the Electricity Council's Annual Report and Accounts, covering generation, transmission and distribution, for 1988/89, a total of 230 billion kilowatt-hours were supplied and the total revenue was £11.8 billion, so the average cost was 5.1 pence a kilowatt-hour. The author in a London suburb took 4264 kilowatt-hours and paid £278, including a standing charge of £35, so his average cost was 6.5 pence a kilowatt-hour, i.e. 27% higher than the national average cost. Even so, the author, as a small consumer far from the largest generating stations, was probably subsidised by the large industrial consumers in the Midlands and further north.
4) The grid operators' model, which was developed over the 60 years or so of its use until the supply industry was privatised, is described in *Modern Power Station Practice* (3rd edition), Volume L, Chapter 1, British Electricity International.
5) 'Through tariffs to dynamic pricing' by R.A. Peddie (consultant) at the Fifth International Conference on Metering and Tariffs for Electricity Supply at the Institution of Electrical Engineers. GP. No. 277, 1987.
6) The expression 'point of supply' is to be applied strictly. If consumers are supplied at different voltages, they are separated electrically by at least one transformer and so are not supplied at the same point, although their premises may be adjacent.
7) By numerically integrating the product of the demand and the dearness over that period.
8) E.g. in Britain the days between Christmas and the New Year, which are more like holidays than normal working days.

9) *Vide* the IEE paper, 'Through tariffs to dynamic pricing', cited in Footnote 5.
10) See Footnote 7 in Chapter Four, particularly the discussion on 15 February 1961. The principal critics were the then chief accountant of the CEGB and the author of 'Electrical Engineering Economics', published in 1936.
11) 'The Control of British Telecom's Prices', July 1988, Oftel. In the end Oftel did not persist with the special tariff, but it is instructive to dissect the mistaken reasoning which underlay it.
12) Oftel tried to estimate that value statistically, but gave it up in preference to a special tariff for small users, without stating what the difficulties were.

APPENDIX 1 – THE DEARNESS MODELS

The first model is represented by the simple equation:

$$a = A = \Sigma C/\Sigma Y$$

where $a = A$ is the dearness (conveniently expressed in £/MWh rather than p/kWh), which is constant in this model,

C is the economic cost of any station (£),

ΣC is the total economic cost of all the stations (£),

Y is the annual output of any station (MWh)

and ΣY is the total annual output of all the stations (MWh).

Whence the dearness worth, W, of any station (£) is given by:

$$W = A.Y$$

The corresponding equation of the second model is:

$$a = B.x$$

where B is the constant parameter (£/MWh),

x is the ratio of the total output at any instant to the average total output,
a now varies from instant to instant in proportion to x.

Whence $W = B.Z$ where $Z = \int_0^T x.y.dt$

y is the output of the station at that instant (MW),
t is the time of the instant from the beginning of the year (hours),
and T = 8760 hours in a non-Leap year or 8784 in a Leap year.

Z can calculated with sufficient accuracy from what is known as an 'output-duration diagram'. That is a diagram in which, against any magnitude of the total output, is plotted the sum of the periods during which the total output is equal to or greater than that magnitude, as shown in Figure 12. The area

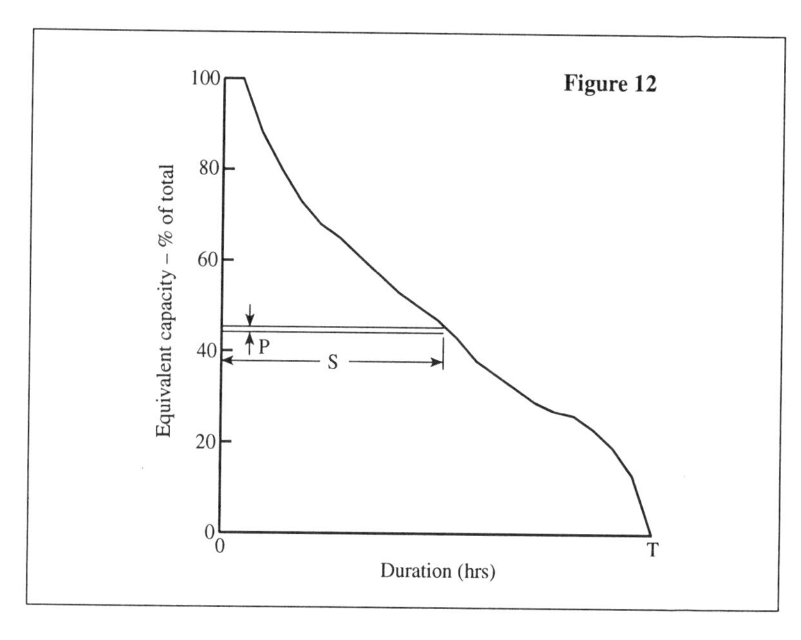

under the curve then represents the total annual output (MWh).

The output of any generator can be represented on the diagram as a thin stratum stretching from the vertical axis to the total output line, whose area is equal to the annual output of that generator. The aim of co-ordinated generation is to operate every generator at its full capacity whenever it is connected to the grid, so ideally the strata of all the generators should be perfectly horizontal, of uniform thickness and arranged in the merit order, starting from the head of the list at the bottom of the diagram.

To allow for some of the inevitable imperfections in the co-ordination, the 'equivalent' capacity, P, of every generator can be calculated as the ratio of its annual output to the sum, S, of the periods during which it is connected to the grid. Then the diagram can be constructed by laying down the outputs of the generators as strata of uniform thickness equal to their equivalent capacities, one on top of the other, in order of decreasing duration from the bottom to the top of the diagram.

When the diagram is so constructed, Z for any generator is given by:

$$Z = P.\int_0^S x.dt$$

the calculation of which is straightforward. And of course Z for any station is the sum of the Zs for its generators. (The calculations for Figure 8 made a further adjustment for the obliquities and tapers of the strata resulting from the imperfections of co-ordination, but that is not essential to understanding the principles of the model.)

In both models, and in any more elaborate model, the difference between the dearness worth of any station and its economic cost is of course $W - C$, which is positive for some stations and negative for others. Equating ΣW with ΣC leads directly to the equation of the first model. Doing the same for the second model leads to:

$$B = \Sigma C / \Sigma Z$$

It is obvious from Figures 9 and 10 that the second model fits the data much better than the first. In statistical theory, the closeness of the fit is measured by what is called the 'standard deviation' of the differences between the two curves in the figures, which is derived from the 'variance'. The quantities plotted in the figures are not W and C but w and c, where $w = W/P$ and $c = C/P$. The variance, V, and the standard deviation, D, are given by:

$$V = \Sigma\{(w - c)^2.P\} \quad \text{and} \quad D = \sqrt{V/\Sigma C}$$

The standard deviation of the first model works out as 3.7% of the total economic cost and in the second model it is 1.1%, showing that the second model fits the data more than three times better than the first.

A third model

For any further improvement, two components are necessary, but a model incorporating A and B is hardly any better than the second model with B alone. What is needed is a component which represents that part of the total standing cost which has to be borne in order to provide emough capacity to meet the highest peak of the total demand, after subtracting the cost which is incurred for the purpose of reducing the running cost. Such a model cannot have a simple equation for the dearness, but the dearness worth of any station can be expressed as:

$$W = U.P + B.Z \quad \text{or} \quad w = U + B.z$$

where U is the second constant parameter
and $z = Z/P$

The two parameters can be calculated by the 'least squares' mcthod, which minimises the variance. In this model the variance is given by:

$$V = \Sigma(w^2.P) - 2.\Sigma(w.c.P) + \Sigma(c^2.P)$$
$$= U^2.\Sigma P + 2.U.B.\Sigma(z.P) + B^2.\Sigma(z^2.P)$$
$$- 2.\{U.\Sigma(c.P) - B.\Sigma(c.z.P)\} + \Sigma(c^2.P)$$

which is a minimum when

$$\partial V/\partial U = 2.\{U.\Sigma P + B.\Sigma(z.P) - \Sigma(c.P)\} = 0$$

and $$\partial V/\partial B = 2.\{U.\Sigma(z.P) + B.\Sigma(z^2.P) - \Sigma(z.c.P)\} = 0$$

Whence B = £32/MWh, U = £17,000/MW and the standard deviation is 0.32%. Such a model could be developed to forecast the dearness, as outlined in Appendix 2. However both the parameters and the standard deviation are highly sensitive to small changes in the cost data which do not affect the single parameter of the second model or its standard deviation. For that reason it would not be worth developing a third model until the consumers have responded to tariffs derived from the more robust, second model.

APPENDIX 2 - FORECASTING THE TOTAL DEMAND

In the first model, the total demand does not have to be forecast in any detail because the dearness is constant.

In the second model, the dearness is the product of the parameter (4.1 p/kWh) and the forecast total demand, expressed as a ratio of the average total demand during the year. The magnitude of that average affects the total cost of supplying all the consumers and it may affect the parameter, but it does not affect the total demand so expressed. That measure of the total demand fluctuates cyclically and randomly but it does not grow or decline. It can therefore be analysed by a modified form of Fourier analysis to separate its cyclical components, which can be forecast, from its random residue, which cannot.

The analysis represents the average of the total demand on the mth day of the year, X_m, as a Fourier series of the general form:

$$X_m = 1 + X_{m1}.sin(p.t_m - \phi_{m1}) + X_{m2}.sin(2.p.t_m - \phi_{m2}) + ... + X_{mr}$$

where t_m is the time to noon on the mth day from the beginning of the year (hours),
i.e. $t_m = 24.m - 12$,
$p = 2.\pi/8760$ or $2.\pi/8784$ in a Leap year,
X_{m1}, X_{m2}, ... are the amplitudes of the annual components,
ϕ_{m1}, ϕ_{m2}, ... are their respective phase angles,
X_{mr} is the residual or random element of X_m.

And the total demand, x_m, during each day is represented as a Fourier series of the form:

$$x_m = X_m + x_{m1}.sin(q.t - \varphi_{m1}) + x_{m2}.sin(q.t - \varphi_{m2}) + ... + x_{mr}$$

where t is the time from noon on that day (hours),
$q = 2.\pi/24$
x_{m1}, x_{m2}, ... are the ampliitudes of the daily components,
φ_{m1}, φ_{m2}, ... are their respective phase angles
and x_{mr} is the residual or random element of x_m.

X_{m1}, X_{m2}, ... and ϕ_{m1}, ϕ_{m2}, ... are of course calculated to minimise the variance of X_{mr} and x_{m1}, x_{m2}, ... and φ_{m1}, φ_{m2}, ... are calculated to minimise the variance of x_{mr}. However the predictability of the components then depends on their continuity over a number of years. In the absence of other indications, the forecast magnitude or amplitude and phase angle of any component are simply its averages in past years, which must be small if the actual values fluctuate greatly. It is to be expected that the first few components will be highly predictable and the later components much less so, but the analysis can be virtually automatic, unless and until evidence of a trend in one direction or another emerges.

The problem of distinguishing between random fluctuations and trends arises whenever data from past events are used to make forecasts, for example forecasts of the flows of reagents in a chemical process plant, the weather or the demand for a public service. The important additional feature in the latter case is that the demand is affected by the prices and methods of charging, which do not come into the other two examples. That factor must severely limit the accuracy of any forecast until the consumers have adjusted their demands to take their best advantage from the tariffs, but thereafter the accuracy may be expected to improve.

In the third model, the dearness is no longer simply related to the total demand. Its capacity component (with the parameter U) requires a different approach.

During periods of low demand, all but the base-load stations (with the lowest running-cost parameters) are either shut down or running at much less than their full capacities. Only when the total demand is at or near its highest peak do all or nearly all the stations contribute to meeting the total demand approximately in proportion to their capacities. It follows that, if the capacity component is to account for the capacity cost of every station, it should be zero at all times except when the total demand is at or near its highest peak. It should then rise rapidly to a large magnitude in an impulse of short duration.

However, although it can be confidently forecast that the highest peak of the total demand will occur on a working day in the winter (in Britain) and that it is much more likely to occur between 8 and 11 a.m. or between 4.30 and 6 p.m. than at other times, it is impossible to predict precisely when it will occur, and it will remain impossible even when the consumers have fully adjusted their demands to take account of the new tariffs.

The reason is that the timing of the highest peak is determined by a combination of factors which are inherently unpredictable, or which can be predicted only a few hours or a few days in advance. They are principally the weather and the uncoordinated actions of the consumers in increasing and reducing their demands from day to day and from instant to instant, to suit their own purposes. The appropriate technique in those circumstances is to forecast the probability that the highest peak will occur on any particular day and during any particular period during the day, and to use those probabilities as the bases for charging for the capacity component of the cost of supply.

Such forecasts involve no more than an analysis of the frequencies of occurrences of the highest peaks from the records of past years. However that line of research will not be pursued here for the reasons given in Appendix 1 (page 231).

7

REGULATING MONOPOLIES AND OLIGOPOLIES

In Great Britain most of the commercial public services are now in the private sector, the most notable exception being the Royal Mail. Before privatisation they were monopolies and, although one of the aims of privatisation was to introduce an element of competition wherever possible, nevertheless seven of them are still monopolies. They are the water, sewage, gas and electricity services to or from the customers' premises, the gas transportation network, the electricity interconnection and bulk transmission network and the railway tracks.

They are natural or technological monopolies because it would be physically impracticable or absurdly expensive to have more than one independent provider. Some competition is feasible further down the line and has been introduced, for example, in extracting natural gas and generating electricity, but the degree of competition is inevitably restricted and such situations are best described as oligopolies. In many countries there is still no competition in those sectors.

For postal and telephone facilities, more than one provider can feasibly reach the customers' premises. Nowadays fibre-optic cable networks can offer alternative telephone and data transmission services, but again the degree of competition is restricted and it is not certain that they will survive in the long term in Britain without the restrictive legislation which set them up in competition[1]. In other parts of an economy the scope for genuine competition is of course much greater, but to make a logical progression from the previous chapter, this chapter begins at the monopoly end of the spectrum.

A publicly owned monopoly can be required to take instructions on its prices and methods of charging, if they are sufficiently explicit, but it has to rely on the public-service motive to induce its management and work force to seek and adopt ways of reducing the costs and improving the quality of its service. Private ownership invokes the motive of self-interest in the shareholders and management and – potentially at least – in the work force, whether shareholders or not, via profit-sharing. But there is then the problem of harnessing that motive effectively to reducing the costs and improving the quality of the service, while restraining the supplier from taking advantage of the customers' dependence on the service to increase its profits at their expense. To some extent, the utilities are in competition with one another, for example gas or electricity can be used for heating buildings and letters can be sent instead of telephoning, but those degrees of competition are not enough, by themselves, to restrain the suppliers from taking advantage of that dependence.

The pros and cons of public versus private ownership of such monopolies depend on one's view of human nature, about which technologists know as much as (though not more than) other people. Adequate motivation is an essential ingredient for prosperity from technology; there is no point in a society putting its monopolies into private ownership if the self-interest of the shareholders and work forces cannot be reconciled with the interests of that society. Some statutory regulation is essential.

The conventional wisdom is that there are broadly two methods[2] of regulating the profits of a monopoly in private ownership. The first, which is best known from its use in the USA, is to limit the so-called 'rate of return'[3] on the capital used for the regulated services. The second method is to 'cap' the prices directly, in other words to place a ceiling on the supplier's prices . The problem is that both methods undermine the inducement to become more efficient.

Under the first method, internally unpopular cost-cutting programmes tend to be shelved because the benefits, which might be shared by the shareholders, the management and the work force, are forfeited as soon as the permitted rate of return is exceeded. On the other hand, if the permitted rate is higher than the rate at which the supplier can raise new capital by borrowing or issuing new shares, it will be tempted to invest in plant and equipment more heavily than necessary for an economic service, so as to widen the capital base to which the permitted rate can be applied in calculating the permitted profit.

So the regulator has the delicate task of setting the permitted rate of return between close limits. It must be high enough to induce the supplier to make innovations, which always run some risk of failure, but not so high that the supplier invests extravagantly. Rate-of-return regulation can prevent a supplier from making profits which are seen by the public as unacceptably high, but indirect exploitation, by inefficient working practices, tends to go unchecked.

At the time of writing, the regulators of the privatised public services in Britain generally favour the second method – price capping – for those reasons, albeit with some reservations. They normally employ a so-called 'RPI-X' formula under which the average of the prices charged by the supplier must not rise more rapidly than the Retail Price Index minus a fixed percentage (the 'X' in the formula[4]). Sometimes X is different for different parts of the supplier's business and sometimes – for example to enable the supplier to meet new quality standards – it is negative, thus permitting the supplier to increase its prices by more than the rate of monetary inflation.

The principal disadvantage of price capping is that, in fixing the permitted price levels, the regulators have to rely on their inevitably uncertain estimates of the likely profits at those prices. Utilities tend to be capital intensive and so their profits depend on how intensively they can use their fixed assets, which depend in turn on the state of the economy, the prosperity of their customers' businesses, the weather and other such uncertain factors. Those uncertainties can cause large errors in the profit estimates.

Another disadvantage is that only the suppliers can know, in detail at first hand, the full range and size of the opportunities for economies in their industries. The regulators' information is always at second hand. Also the little word 'average' has come to conceal some arbitrary weightings in the calculation[5].

To mitigate those disadvantages, the regulations are reviewed periodically, typically at five-year intervals but more often if there are strong public complaints of monopoly profits. Under price capping the suppliers can retain their profits at least until the next review, which the regulators claim gives them enough room to manoeuvre, but five years is not very long in a capital-intensive industry and a supplier's fears of the outcome of the next review are likely to inhibit its pursuit of economies for profit meanwhile, especially towards the end of the period.

So great are the disadvantages of those two conventional methods that it is commonly concluded that competition is the only effective remedy and that even a little competition is better than none. Thus in privatising the electricity supply industry in 1989, the British Government went so far as to break up the national generating capacity into a few large companies, sacrificing the smooth working of co-ordinated generation described in Chapter Six (page 201) in the process, just in order to create a modicum of competition.

However, whilst competition can often restrain suppliers from exploiting their customers, it also has some disadvantages. *The new approach in this chapter is to consider which features of competition are desirable and should, if possible, be simulated and which are undesirable and should be omitted in formulating a regulation method for monopolies.*

THE MERITS AND DISADVANTAGES OF COMPETITION

Competition for survival is said to be universal among all the biological species, and it is probably the main driving force of evolution, but *homo sapiens*, among other species, is a social animal and we owe much of our success to our ability and habit of elaborate co-operation between large numbers of individuals, particularly in large-scale applications of technology. In that sphere at least, competition is not an end in itself. It is one means, but not the only means, of inducing the owners of and workers in a commercial supplier to economise and to pass on the benefits of the economies to their customers.

The best known disadvantage is that, for the competition to be fully effective, the suppliers must be too numerous to have any considerable leverage on their prices, which are supposed to be governed by market forces. To be numerous the suppliers must be comparatively small and so inevitably forego some of the important economic advantages of operating on a large scale. If there are only a few suppliers, then they may be able to obtain so-called 'market power', with which they can influence their prices to their advantage at the expense of their customers.

A second, less obvious but important disadvantage arises from the facts that the demand forecasts of competing suppliers are bound to be far more speculative than a monopoly could make and they cannot know for certain what their future individual shares of the market will be. So they require higher prospective rates of return on their capital than a monopoly would need to stay in business. That latter effect soon became apparent when the British electricity generation monopoly was broken up by the Electricity Act 1989.

A third disadvantage, which is not widely understood, is that genuinely competing suppliers can all raise their prices together in response to information which comes to them all at the same time, such as an increase in the price of a common raw material. Such simultaneous price rises are sometimes mistaken for evidence of collusion between the suppliers, but the suppliers do not need to communicate with one another at all to perceive that the lowest price which any of them will be able to maintain in the foreseeable future has risen, and to raise their own prices accordingly, just as they are all obliged to lower their prices when a raw material price decreases. So much is common economic theory, but if the aim is prosperity from technology that immediate response is a disadvantage of competition, in as much as unexpected changes in prices vitiate the long-term calculations of the designers of projects which use the affected goods or services, as explained in Chapter Three (page 60). It would be better for the price variations to be more gradual.

Competition can theoretically inhibit a supplier from favouring one set of customers at the expense of another set. For if a supplier does so – either intentionally or carelessly – an efficient but not necessarily more efficient competitor will be able to undercut that supplier's prices to the second set of customers and so obtain their custom, leaving that supplier with the first set of customers, which it was supplying at a loss, but without the second set from which to recover that loss. However that mechanism is crude and not always effective. It would be better if the supplier could be prevented more directly from discriminating between sets of customers. Such prevention can only be effective to the extent to which the true costs of supplying the customers can be calculated, but that condition also applies when competition is the means of preventing discrimination.

An essential ingredient for effective competition, which is often overlooked, is the publication of the prices at which goods and services are being exchanged. The competing suppliers cannot undercut one anothers' prices effectively without reasonably accurate knowledge of those prices, and the designers of projects which will use those supplies need the information for their design calculations. A regulated monopoly can be required to publish its prices, as a condition of its licence.

Quality is at least as important to the customers as prices but, contrary to what is often unthinkingly assumed, unregulated competition does not lead automatically to high-quality goods and services. A modern economy, being highly dependent on technology, relies on processes which are physically remote from and largely incomprehensible to most of the customers. The appearance of the products seldom enables the unaided customers to discern their true quality at the time of purchase or supply. Stringent regulations are essential, at least to label the products accurately and informatively and often to require compliance with quality standards, and they must be monitored and enforced. It is arguable that less legislation and fewer inspectors are required to regulate effectively the service quality of a monopoly than of competing suppliers.

In some sectors of an economy, such as food, clothing and jewellery retailing, great variety is also very important. Thus a supermarket may offer dozens of varieties of hundreds of different products for the delectation of its customers. It is not suggested that there is any effective substitute for competition in those sectors. However variety on that scale is not an important factor in the public utilities.

What then are the specific merits of competition in the public utilities which must be set against those disadvantages? Its principal merits are that it gives a supplier the opportunity initially to retain as profit a large part of any economy it can make (by cost reductions or by increasing its turnover more rapidly than its costs or both), but that that retention is eroded when the competing suppliers catch up, by adopting the same or other economies. Eventually all the competitors must drop their prices to stay in business, and in so doing they pass on those economies to the customers.

The disadvantage of small-scale operation disappears automatically in any monopoly and the problems of demand forecasting and quality control are lessened. Concealment of prices can be prohibited if its undesirability is recognised. So the main criteria for judging one monopoly regulation method against another are the extent to which they simulate the foregoing opportunities and pressures of competition and reduce the supplier's scope for price manipulations and discrimination between customers.

ECONOMY-SHARING REGULATION[6]

The source of the inadequacies in both of the foregoing methods of regulating monopoly prices and profits is the absence of any explicit rules for sharing the economies, which the supplier can make from time to time, between the supplier and its customers. Rate-of-return regulation discourages economies from the outset, whilst an RPI-X price-capping formula bears down on the supplier's prices, lightly or heavily depending on the value of X, but without full cognisance of the supplier's current opportunities for economies, which may vary widely – and sometimes rapidly – from time to time. The remedy must be to formulate a set of rules which simulates the pressures of competition whilst avoiding its four main disadvantages: small-scale operation, speculative demand forecasts, price manipulations and unexpected price changes.

The details are less important than the principle, which is to make the supplier's permitted profitability depend on the benefits obtained by its customers from its price reductions. However there is no point in making the rules more complicated than necessary and the rules which follow are the simplest that can be devised to fulfil that principle. It is convenient to express them as a formula[7] with three components. All the quantities in the formula are real, in the sense of having been adjusted for inflation.

There are two feasible measures of profitability, namely (1) the real profit calculated by Adjusted Historical Accounting as described in Chapter Five, or (2) the total of the distributed equity dividends. Each has its merits and some disadvantages. For convenience let us first apply that principle to regulating the supplier's profits and then go on to apply it to the supplier's dividends – for consideration as an alternative.

The first component of the formula is the permitted base level of profit which is independent of improvements, in other words the profit which the supplier is allowed to make without reducing its (real) prices. It is a fixed percentage, termed the 'base rate', of the equity capital. It allows the supplier to raise enough capital to continue in business when no economies are possible. The real rate might be between one and five per cent a year, perhaps three per cent[8], bearing in mind that monopolies are inherently low-risk enterprises, especially if they do not try to economise.

The second component determines what proportion of the economies made by the supplier from time to time may be retained as additional profit and what proportion must be passed on to the customers in (real) price reductions, as the condition of retaining that additional profit. The 'initial benefit' obtained by an individual customer from the price reductions in any year is defined as the difference between what that customer was charged and what they would have been charged for the same service in the previous year. It is a putative benefit in that it is calculated from this year's and last year's prices and this year's consumption or utilisation of the service, without regard to last year's consumption or utilisation.

Customers may also gain or lose from using the service more or less this year than last year, but those gains or losses are not relevant because they are not the direct outcome of the supplier's price reductions. Moreover, customers continue to benefit from the price reductions in earlier years for as long as the lower prices are maintained and they remain customers, and new customers after the reductions also participate in those benefits.

The (total) customers' initial benefit in any year is simply the sum of the initial benefits, so calculated, of all the individual customers in that year. The formula makes the permitted additional profit equal to the customers' initial benefit in that year, multiplied by a factor, plus a proportion of the permitted additional profit in the previous year. For example, if the factor is 3.0, the proportion 75% and the customers' initial benefit this year £100 million, the permitted additional profit this year is £300 million plus 75% of last year's permitted additional profit. If that was £200 million, then the permitted additional profit this year is £300 million + 75% of £200 million = £450 million – and so on. The factor is the means of inducing the supplier to share its economies with its customers straight away; let us call it the 'sharing factor'. The proportion allows the supplier to continue deriving some profit from those economies in later years; we may call it the 'continuance proportion'.

Those calculations are retrospective and free from the necessity of subjective judgement on the part of the regulator. But of course the supplier has to work from its forecast business volume and intended economies and judge how to pitch its prices so as to maximise its profit. It must lower its prices in order to be allowed to increase its profit but the lower prices will then reduce the profit it can actually make. That situation is close to the competitive situation in which a supplier must lower its prices in order to gain business which would otherwise go to its competitors, but again the lower prices reduce the actual profit. The principal difference is that the formula is logical and predictable, whereas the behaviour of competitors is often unpredictable and sometimes illogical.

The third component in the formula, which also features in the conventional methods, caters for the impossibility of exact forecasting. As long as the profit in any year is no greater than the permitted profit (the sum of all three components), the deficit (that is the permitted profit minus the actual profit) is carried forward to the permitted profit in the following year. So the supplier need not strive to pitch its prices to make exactly the full permitted profit every year. It can pitch them a little lower, knowing that if it makes less than the permitted profit in one year, it will be permitted to retain a correspondingly larger profit in the next year. Thus the supplier can avoid the risk of pitching its prices too high.

It has to be expected that sometimes the supplier may wish to raise its (real) prices, or may be obliged to do so by external events, thus generating a negative customers' initial benefit, which will reduce the permitted additional profit. The intention of the method is that the supplier shall normally accept the resulting lower profit, just as it would under competition.

The supplier may also – by accident or intentionally – make more than the permitted profit in any year. In that situation, the excess profit, plus interest at a specified rate, is carried forward to the next year as a negative profit deficit. That interest constitutes the necessary penalty to deter the supplier from making more than the permitted profit intentionally and to encourage it to take care not to do so accidentally. Some interest might also be added to the positive deficit when the actual profit is less than the permitted profit, but that is not an essential feature of economy-sharing regulation.

A numerical example

The table below shows how a hypothetical supplier might fare under the formula with a base rate of 3% a year, a sharing factor of 3.0 and a continuance proportion of 75%, when its business expands and its equity capital base grows at a constant (real) 5% a year for nine years. The supplier is assumed to have held its (real) prices constant over the years up to and including the first year of the example (when its turnover is £4,000 million), so its profit in the first year is down to 3% of its equity capital (£10,000 million) at the beginning of the year.

CUSTOMERS' BENEFITS & SUPPLIER'S PROFITS

Yr	Supplier's				Customers' benefit		Profit		Rate
	eqty. cap. £m	ex-pend. £m	price red. %	turn-over £m	init. £m	total £m	per-mittd. £m	act-ual £m	on cap. %
1	10000	3700	0	4000	0	0	300	300	3.0
2	10500	3110	4.8	4000	200	200	920	890	8.5
3	11030	2610	4.8	4000	200	410	1420	1390	12.6
4	11580	2190	5.2	3980	220	650	1820	1790	15.4
5	12160	1840	5.3	3960	220	910	2150	2120	17.4
6	12760	1930	3.2	4020	130	1080	2120	2090	16.4
7	13400	2030	3.0	4100	130	1260	2100	2070	15.4
8	14070	2130	2.9	4180	130	1450	2080	2040	14.5
9	14780	2240	2.8	4260	120	1650	2060	2020	13.7
10	15510	2350	2.7	4360	120	1850	2043	2010	12.9
Averages (over years 2–10)			3.8			1050		1820	14.1
15	15510	2350	1.9	3930	70	2270	1610	1580	10.2
20	15510	2350	1.5	3630	50	2580	1300	1270	8.2

The supplier is assumed to make economies in each of the next four years, thereby reducing its expenditure to the figures in the third column. But in order to increase its profit above 3%, the supplier must also reduce its prices. It is not obliged to reduce all its prices equally, but for simplicity it is assumed to do so in this hypothetical example.

The successive reductions in the fourth column permit the supplier to build up its profits to 17.4% of its equity capital in the fifth year, as shown in the right-hand column. Thereafter the supplier is assumed to make no further economies, so its expenditure increases by 5% each year but, as long as those earlier economies are not undermined, the supplier can continue to profit from them. However it must continue to reduce its prices, as shown in the table, in order to prevent its permitted profit from falling rather rapidly. At best, with a small allowance for forecasting errors, it can keep its profits up to the series in the right-hand column, which falls to 12.9% in the tenth year. The formula and the calculations are given in the appendix.

Figure 13 shows how the supplier's actual profit (always slightly less than its permitted profit) increases to £2,120 million in the fifth year and then declines to £2,010 million in the ninth year, averaging £1,820 million a year over the nine years. The figure also shows how the customers' total benefit builds up. In any year it comprises their initial benefit from the supplier's price reductions in that year plus their continuing benefit from its reductions in previous years. It is the difference between what they paid for their services and what they would have been charged in the absence of any price reductions since the first year. It increases to £1,850 million in the ninth year, averaging £1,050 million a year over the nine years.

Those benefits and profits are the outcome of the supplier's economies and price reductions and the values of the parameters in the formula. Some readers may feel that in this example the supplier has been allowed to make too large profits for too long and that, if the parameters had indeed been set at those levels, they would have to be reviewed after five years, if not sooner. Or they might grant that, under

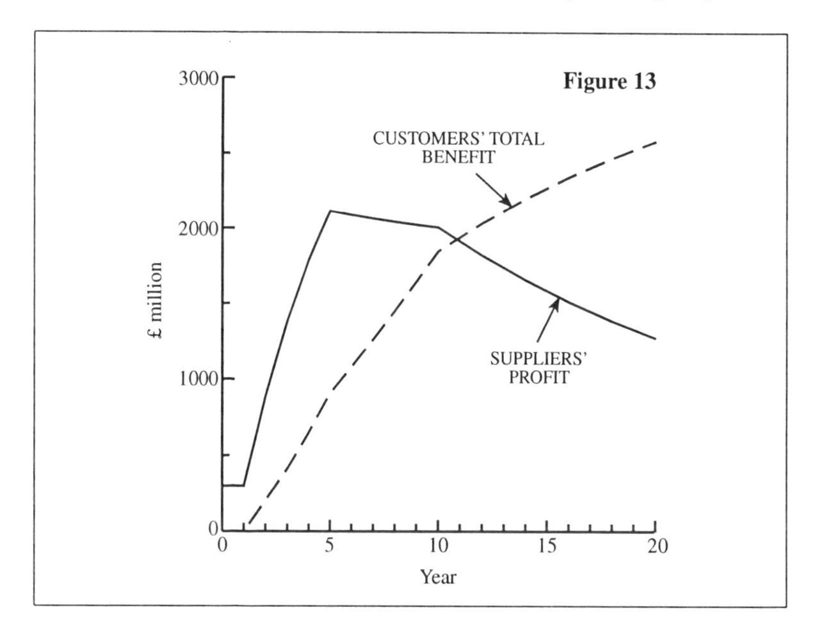

competition, such expansion and economies often yield large profits in the first few years but that a review would certainly be required after ten years. However, before considering how much weight should be given to such feelings, it is worth seeing what might happen if the initial values of the parameters were retained indefinitely.

In the example, the supplier is assumed to cease expanding its services after the tenth year and to make no further economies. Nevertheless, it continues to profit from its earlier efforts, albeit at a gradually declining rate – down to £1,270 million (8.2% of its equity capital) in the twentieth year. Meanwhile the customers continue to receive marginal price reductions which add to their total benefit. It overtakes the supplier's profit in the eleventh year and and builds up to £2,580 million in the twentieth year. In the long run, the whole benefit of the supplier's efforts in the early years is transferred on to its customers.

Thus the supplier is induced, in the interest of its shareholders, to keep trying to economise and passing on part of its savings to its customers; hence the name 'economy-sharing regulation'. If it fails to reduce its costs, in relation to its turnover, or it does not pass on part of the reductions to its customers, then its permitted profits will fall, inevitably, to the base level. If it makes large economies, it can retain large profits for a time, but not indefinitely. It must continue to cut its prices every year, irrespective of further economies, as long as its profits are greater than the permitted base profit.

An additional feature of economy-sharing regulation is that it induces the supplier to raise its prices less after a raw-material price increase than a set of competing suppliers in the same situation, or even to continue lowering its prices. And it induces the supplier to lower them less, or less rapidly, after a raw-material price decrease. In both cases the supplier is induced to smooth out its own price changes, thus reducing the losses of its customers in two critical sets of circumstances, mentioned in Chapter Three (page 60). Those are: (1) when a customer has become committed to a project which depends on a supply from the monopoly, on the basis of calculations at the old *lower* price, and (2) when a customer has decided against such a project, or has become committed to a project which provides that supply from another source, on the basis of calculations at the old *higher* price.

Tariffs of monopoly suppliers

To make its best contribution to prosperity, a monopoly supplier must adopt a method of charging which is rational and non-discriminatory. If a rational method, such as has been proposed for public electricity supply in the previous chapter, has been established, it might be made mandatory, but it is not essential for the regulator to specify the method because the economy-sharing formula offers an independent test of those qualities. Indeed, in as much as it would be difficult, if not impossible, for the regulator to supervise the supplier's application of a mandatory method in the necessary detail, it would probably be better for the regulator not to attempt to do so but rather to rely on the test. He or she would be obliged to do so in any case in an industry for which no rational method of charging has been worked out.

The test is that, with two necessary exceptions, it must be possible, from the supplier's published tariffs and the records of any customer's consumption or utilisation of the service, to calculate unequivocally how much that customer would have been charged for any year's consumption or utilisation under the previous year's tariffs.

One necessary exception is when the supplier is obliged to incur additional costs to comply with new legislation, for example on new mandatory quality standards or obligations to counter the effects of pollution not previously the responsibility of that supplier. The crucial feature, which distinguishes such occasions from the occasions when the supplier is expected to absorb the additional costs, is the incidence of new legislation.

Under any method of regulation, natural justice would require the rules to allow a supplier to meet such newly imposed obligations without loss of profit, and the contribution of the proposed method to prosperity would be diminished if it failed in that respect. The parameters of the formula would have to be adjusted, but the adjustment would be no more difficult to assess than the corresponding adjustment to a price capping formula.

The other exception arises when the supplier introduces a new type of service – new, that is, in the sense that it is recognised by the regulator to be unavailable under any existing tariff. In the first year of a new service there are no prices for the previous year on which to calculate the customers' initial benefit, but in as much as a new service normally has small beginnings, the supplier's initial prices must be low enough to attract the new business. So the supplier's total profits from that source in the first year cannot be excessive. If the supplier is allowed to retain its actual profit – using normal cost-accounting methods – in the first year or two of a new service, thereafter the formula will automatically begin to transfer the benefits to the customers.

Normally the supplier would be expected to present the calculations to the regulator in pursuit of its claim for permission to retain its profit, but with the stipulation that the calculations must be explicit, so that the regulator can check them for any selected sample of customers. Non-discrimination could then be induced by requiring that the charge for one year's supply at the previous year's prices must be the lowest which can be calculated under any tariff available in the previous year.

It may perhaps appear at first sight that the calculations would be too numerous and complicated to be practical, but that appearance is no more than a carry-over from the days before computers performed all the calculations for billing the customers. In those days, when armies of clerks did the calculations by hand, it would indeed have been too expensive to require every customer's bill to be calculated twice, once at the current year's prices and again at the previous year's prices, then to take the difference between them and finally to add up all the differences for millions of customers. And the results might not have been sufficiently reliable. Economy-sharing regulation would not have been a practical proposition in those days and would not have been put forward for that reason.

But nowadays the additional calculations do not present a serious problem. Last year's prices can be retained in the computer memories alongside this year's prices, the two calculations for each customer performed in succession and the differences summated in a running total. The only exceptions are when the supplier introduces a new type of service which is recognised as such by the regulator, and in those cases the second set of calculations are not required, as already mentioned.

Suppose now that the supplier can calculate that it is charging one set of customers considerably more in relation to its costs of supplying them than another set. Or, to put it another way, suppose the supplier knows that it is making a large profit from supplying one set of customers but a very small profit – or even a loss – from supplying another set. It will also be aware that, in general and other things being equal, reducing its prices to any customers will induce them to increase their consumption or utilisation of the service, whereas increasing its prices will have the opposite effect.

It follows that, if the supplier reduces its prices to the former set of customers, their consumption or utilisation will probably increase, but (provided the reduction is not too great) its costs of supplying them will increase by less than its revenue from that source. That is to say its profits will probably rise; moreover the price reductions will permit those larger profits to be retained under the formula. On the other hand, if the supplier increases its prices to those customers, their consumption or utilisation may fall, so reducing its profits from that source and also its permitted profits overall. And vice versa for the latter set of customers.

In other words the supplier will find that the long-term interests of its shareholders are best served by aligning its charges to any set of customers as closely as possible to the costs of supplying those customers, that is to say by setting its prices to obtain the same pro rata profits from all its customers. Its ability to do so will depend on it working out a rational system of prices and methods of charging, but its first-hand information and large financial resources will always put the supplier in a far better position to conduct the necessary research for that policy than the regulator can hope to emulate.

That is how economy-sharing regulation simulates one of the beneficial effects of competition, but under competition the supplier is not under the same continuous pressure to make price reductions in order to retain the profits arising from rationalising its prices. So with economy-sharing regulation in place, the regulator need not have any particular level of prices or method of charging in mind in scrutinising the supplier's applications to retain its profits.

That property of inducing the supplier to make its tariffs rational and non-discriminatory as far as it can do so, without any interference from the regulator, is a feature of economy-sharing regulation which the other methods do not have.

Setting the parameters

The formula has three parameters, namely the base profit rate (3% in the example), the sharing factor (3.0) and the continuance proportion (75%). Three parameters are enough for the purpose of inducing any monopoly in private ownership to economise in the interests of its customers. However questions such as: is 3% a high enough base rate to attract enough equity capital to keep the business barely running?; could it be lower?; would a larger or smaller sharing factor stimulate or discourage the work force via employee-share-ownership or other profit-sharing?; and would a smaller continuance proportion make the managers more alert or more despondent?; must depend on the state of the industry and the economy in which it is to operate and on the natures – perhaps one should say the psychology – of the shareholders, managers and work force. In those respects, economy-sharing regulation requires as much thought before it is introduced as either of the other methods.

The base rate is intended to allow the supplier to raise enough equity capital to continue in business when no economies are possible. It is intended to be based on the adjusted historic total of the ordinary shareholders' funds, including receipts from successive issues and reserves accumulated from past earnings. The supplier is of course free to raise separate, non-equity capital by borrowing and/or other means and to deduct the interest and/or fixed-rate dividends on that capital in calculating its profit, but the base rate does not correspond to the permitted rate on the supplier's whole capital base under rate-of-return regulation. That rate had to match the earnings yield required by the market on new equity issues because it had to provide a return on risk capital, but under economy-sharing regulation the rewards for taking risks come from the second element in the formula.

The sharing factor simulates the opportunity under competition to retain the profits from any economy in the early years and the continuance proportion simulates the erosion of those profits as the competition catches up. It would therefore be appropriate to derive those two parameters from studies of the financial constraints under which genuinely competing companies have to operate. However the choices of those two parameters and the review periods are much less critical than in the other methods because their purpose is to regulate the sharing of the benefits of the supplier's economies between its shareholders and its customers in the short term and the medium term; they do not affect the eventual outcome, even if there is no review.

Figure 14 shows the effects in the hypothetical example of reducing the sharing factor from 3.0 to 2.0, while increasing the continuance proportion from 75% to 84%. The supplier's profit then peaks at £1,930 million instead of £2,110 million and the customers' total benefit overtakes the supplier's profit in the tenth instead of the eleventh year. But the supplier's profit and consumers' total benefit in the twentieth year are not affected. The profit is bound to fall back to the 3% base rate (£470 million) and the consumers' total benefit is bound to

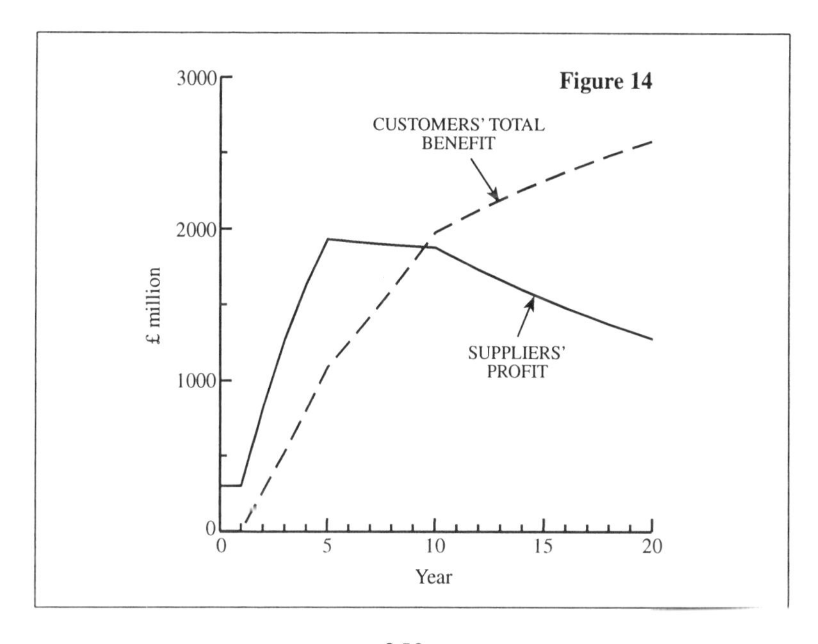

rise to £3,850 million a year in the end, whatever the settings of the sharing factor and the continuance proportion may be.

That eventual, total transfer of the benefits of the supplier's economies to its customers is the most important feature of economy-sharing regulation which the other methods lack.

Once the parameters have been decided, the formula poses no great computational difficulties for the supplier in working out its best strategy for maximising its profit from any economies (including economies from expansion of turnover) it expects to make. Indeed an important advantage over price capping is that the supplier can do so without having to wonder how tightly the regulator will cap its prices after the next review and how large a profit it dare make meanwhile. Changes in the formula can be limited in advance and (in the absence of new legislation affecting the supplier's costs) can be at less frequent intervals than five years, because the formula caters for a wide variety of unforeseen developments, including growth and contraction of turnover, without any changes to its structure or parameters.

It remains to pick up a loose end which was left dangling in Chapter Three. The resource costs of any monopoly include the necessary profit on capital investments to take account of the financial risks, but beyond those profits the additional profits of monopolies in private ownership were classed as a special case in Chapter Three (page 95) to await consideration in the present chapter. We can now perceive that, under the foregoing regulation formula, they are the costs of motivating the shareholders and work forces to economise. The benefit from those economies passes to the customers in the end; hence the customers should provide the profits meanwhile. Their *quid pro quo* is the better motivation of the shareholders and workers to economise in the first place, but without dissipating part of the benefit in what is often wasteful competition. In general the benefits will be proportionately larger for large customers than for small ones; hence the profits should be obtained by pro rata additions to the charges – by the same percentage added to all the customers' bills, which is how the supplier will obtain them when regulated by the formula.

Regulating profits versus regulating dividends

All the regulation methods, including RPI-X, depend on assessments of the profits to ordinary shareholders, but not to the same extent. Inflation accounting apart, any method which sets out to regulate a company's profits is more sensitive than RPI-X to the accounting conventions under which the profits are calculated and so to the scope for creative accounting and other ways of distorting those profits. That greater sensitivity would be a disadvantage of any method, compared with RPI-X.

To overcome that disadvantage, economy-sharing regulation could be applied to the ordinary dividends of the regulated supplier instead of to its profits. Ordinary dividends are easily ascertained and ascertaining the dividends which are distributed – as distinct from the dividends which the supplier may feel it will be prudent to distribute – does not depend on the accounting conventions of the supplier.

The justification of the change would be that ordinary dividends represent the only way in which the supplier can eventually pass on its profits to its ordinary shareholders. Profits and other gains (including capital gains) which are not distributed as dividends generally increase the market value of a company's ordinary shares but they cannot benefit the shareholders directly. The increased market value comes from the prospect of larger dividends in the future, which are normally expected from such undistributed gains, but if the dividends are regulated rather than the profits then the market value will reflect that regulation.

It is important that the quantity to be regulated must be the total sum distributed as dividends (less the basic rate dividends), not the dividend rate; otherwise the supplier could avoid or largely avoid the limitation by issuing bonus shares and paying dividends on them.

An incidental advantage of applying economy-sharing regulation to dividends rather than profits would be that final dividends are not normally distributed until the accounts have been completed. Consequently the total permitted dividends could be calculated before the final dividends are paid to the shareholders and it would be a simple matter to prevent the actual dividends from exceeding the permitted total, thus eliminating the need for a penalty if the profits are accidentally greater than intended. The excess profits would be ploughed back automatically and would not be available to the shareholders until the supplier had reduced its prices sufficiently to warrant their distribution under the formula.

Additional operations

Many of, if not all, the privatised utilities have branched out since privatisation to offer additional products or services which they did not previously provide. The regulators have therefore to rule on whether those new operations should be regulated and, if so, how. For example, the National Grid Company (NGC) now offers several additional services, one of which is telecommunications via fibre-optic cables attached to the earth wires which are slung between the grid towers above the power cables. It is of course much cheaper to run telecommunication cables in that way than to bury them in the ground or erect special poles for them, but they must be fibre-optic cables, which were not developed until the 1980s, because electrical signals in conventional telecommunication cables near to and parallel with power cables would be subject to gross interference from the power currents.

In a public consultation in 1996[9], the Office of Electricity Regulation (OFFER) invited comments on which of NGC's additional services should be included in its price controls and how should their costs be accounted. But OFFER tacitly assumed that it should not try to regulate the prices charged for telecommunications because they are for an entirely different kind of service from NGC's main power transmission business and are sold in a separate market. Yet those telecommunication services rely on the existence of NGC's major tangible fixed asset – the Grid – and they could not be offered at all without the revenue from NGC's main business, which has paid for and is continuing to pay for the construction, extension and maintenance of the Grid. So was NGC exploiting its monopoly in power transmission to inflate its telecommunication revenue and profit outside OFFER's price controls at the expense of those main customers?

That question was ignored in the consultation documents, but OFFER'S subsequent 'valuation of Energis at flotation' in its 1996 price control proposals was more than £600 million lower than the reported market value when Energis came to be floated only a year later. The difference was a windfall gain to the NGC shareholders at the expense of their main customers.

Under economy-sharing regulation on the other hand, none of the services would be price controlled. The question, therefore, is which services should be included in the calculations of the customers' benefit? And the obvious answer is the monopoly services which only NGC can provide. Those services are power transmission and some ancillary services, but not the telecommunication services because they can be obtained from other providers.

Under economy-sharing regulation NGC would be free to set its prices for all its services, including telecommunications (unless they came under the control of the telecommunications regulator), without any direct restrictions. But the revenues from all its businesses would be included in the profit calculations for comparison with its maximum permitted profit in any year. If its telecommunication business made a profit or was sold for a capital gain, NGC would be obliged, in effect, to reduce its power transmission charges in accordance with the formula in order to be allowed to distribute a proportion of that profit or gain from its telecommunication business to its shareholders. Thus the benefits of that exploitation of the Grid would be shared between NGC's shareholders and its main customers, who jointly provide the necessary funds for the Grid. No difficult accounting problems would arise.

The general presumption, applicable to any monopoly, is that all its profits depend, wholly or in part, on its monopoly power; they should therefore be shared in some degree with its main customers because they are tied to the monopoly for the services they require. If any part of a monopoly's business is entirely independent of that power, that part can be detached and taken over by a separate, independent company without loss of profit. A monopoly's willingness to divest itself of a part of its business in that way is a valid test of its genuine belief that that part is indeed wholly independent of its monopoly power.

Economists tend to assume that competition is always a more effective way of curbing the prices and profits of commercial suppliers in the interests of their customers than any formula or rules imposed by legislation and appointed regulators. This section does not claim that the foregoing formula is superior to genuinely effective competition, when great variety is crucial and the suppliers can have so little individual influence on the market that they give up trying to have any and concentrate their efforts on economising to make a profit at the market prices. But the assumption is doubtful as a general principle, applicable in all circumstances, in the light of the disadvantages of competition reviewed earlier in this chapter, and of course it is not relevant to monopolies.

For the natural/technical monopolies the formula is proposed as a decisive improvement on the conventional regulation methods. But what about the oligopolies? Must there be a stark choice between combining (or recombining) the companies in an oligopoly into a monopoly on the one hand or breaking them up into so many units that they will genuinely compete on the other? Can they not be effectively regulated as oligopolies?

REGULATING OLIGOPOLIES

Although they have less power than an unregulated monopoly in regard to prices and methods of charging, the suppliers in an unregulated or ineffectively regulated oligopoly can often raise some prices without losing much business and lower others for the purpose of attracting customers away from other suppliers into an arrangement which is not in the long-term interests of those customers. Or the suppliers may lower their prices below their resource costs during a 'price war', after which they have to charge well above their resource costs to recover their losses. Or one supplier may take advantage of temporary weaknesses among the others to jack up its prices and build up a reserve fund, in preparation for the next price war.

An elaborate theory, known as 'the theory of games' (though it might be more aptly described as 'the theory of commercial warfare') has been developed, largely to help suppliers in oligopolies to manipulate their prices so as to get the better of one another in different situations, with scant regard for the welfare of their customers. Yet the suppliers claim, with some justification since they do compete to a substantial extent, that they must not be required to publish information which they choose to classify as 'commercially confidential'. Consequently any direct regulation of their prices and practices is likely to be severely hampered by a lack of information – and clumsy in consequence. Electricity generation in Britain provides a convenient example of an oligopoly which exhibits the foregoing unwanted features. First some history to set the scene.

A brief history of public electricity supply in Britain

This industry was partly in private and partly in public ownership until 1947, when it was totally nationalised by the Electricity Act of that year. That act set up a Central Electricity Authority, responsible for generation, interconnection of generating stations and bulk transmission, and fourteen Area Boards (twelve in England and Wales and two in Scotland), responsible for distribution in their separate areas.

Subsequently, the four functions of generation, interconnection, transmission and distribution were combined in Scotland, but the original organisation was retained in England and Wales, with the English and Welsh generating stations and Grid vested in the Central Electricity Generating Board (CEGB). A cable link was laid under the English Channel to enable electricity to be exported to or imported from France. Latterly the Grid in England and Wales became a net importer from both France and Scotland.

The privatisation of electricity supply in Britain in 1989 was one of the most controversial measures of any British government in the field of economic policy. It was fiercely attacked and as fiercely defended. The debates were almost all about public versus private ownership of the means of production and distribution of a vital commodity. In those debates one of the main objections to privatisation was that electricity supply is a natural monopoly which, it was argued, could not be adequately regulated under private ownership; it would either be stifled or enabled to exploit the customers' dependence on its services for its own profitability at the customers' expense. The regulation of two previously privatised companies, British Gas and British Telecommunications, was widely felt to be unsatisfactory in that respect.

So the privatisation was geared to break up the monopoly created by the 1947 Act and introduce competition, wherever feasible, throughout the supply industry. In practice that meant competition in generation[10] in England and Wales, at the cost of disrupting the previously developed organisation and methods of co-ordinating the design, construction and operation of the generating stations and the electrical networks. The possibility of privatising the industry without interfering with that organisation or those methods, and then regulating the several monopolies (a privatised Central Electricity Generating Board and twelve privatised Area Boards) so as to induce them to economise and pass on their economies to their customers, was not seriously considered. Perhaps it will be considered in other countries in which electricity supply is still in the hands of monopolies.

Under the Electricity Act 1989, the twelve Area Boards in England and Wales became regional electricity companies (RECs) and they jointly owned the National Grid Company (NGC), in which was vested the Grid, South of the Scottish border, and the pumped-storage stations, described in Chapter Six (page 207), in Wales. That act also divided the generating stations of the CEGB between three newly created generating companies, namely National Power and PowerGen, which between them owned nearly all the coal, oil and gas burning stations, and Nuclear Electric, which owned the nuclear stations. All the companies created by the 1989 Act, except Nuclear Electric, were privatised then.

In Scotland, the four functions of generation, interconnection, bulk transmission and distribution are still combined and OFFER regulates the prices of the Scottish companies. The rest of this section will therefore be concerned only with England and Wales. The table below shows the generating capacities at the privatisation vesting date and at the time of writing. Nuclear Electric was privatised in 1996, except that some older stations were retained under public ownership in Magnox Electric. The RECs divested themselves of their joint ownership of the NGC in the same year and NGC in turn sold the pumped storage stations. A fourth major generating company, Eastern Electric, came into being in 1996 and together the four concerns own three-quarters of the total capacity. The remaining quarter is owned by some 25 minor concerns.

GENERATING CAPACITIES IN ENGLAND AND WALES

Generating company	Capacity at vesting day		Capacity at April 1998	
	Gigawatts[10]	%	Gigawatts	%
National Power	30	48	17	27
PowerGen	19	30	14	22
Nuclear Electric	9	14	7	12
Magnox Electric	–	–	3	5
Pumped storage	2	3	2	3
From France	2	3	2	3
Scotland	1	1	1	2
Eastern Group	–	–	8	13
Others	–	–	7	12
Total	62		61	

Before 1989, the CEGB, as owner of the generating stations and the Grid, issued co-ordinating instructions, based on physical measurements in the stations and its commercial transactions for the supply of fuel and stores, the services of its operating staff and the construction of its stations and the Grid. Those instructions covered the design and construction of new stations and extensions to the Grid, the starting up, loading up, off-loading and shutting down of generating sets, and the scheduling of their repairs and maintenance. Now NGC, on which responsibility for the co-ordination has fallen, is cut off from most of that information.

The outputs of the four large generating companies now immediately cross commercial boundaries, which the Act has created between their stations and the Grid, so NGC must co-ordinate their generation via the market in which those four companies and the other generating suppliers sell electrical energy to the distributors, while allowing, as best it can, for the effects of the energy flowing through the Grid as a result of the direct contracts between the generating companies and some consumers. And, whereas previously the single owner bore the costs of failures to comply with the co-ordinating instructions, due to breakdowns and unexpected problems during the construction, operation, repairs and maintenance of the stations, now those costs must either be recovered from the generating companies in that market or passed on to the consumers in blanket margins which create no inducements for the generating companies to improve their reliability.

The arrangements for that market, known as the 'Pooling and Settlement Arrangements', were negotiated between the bodies which were due to become the main generating companies and the RECs, under an official but self-confessed incompetent chairman[12] and with no participation of the ultimate consumers. The arrangements began by intentionally ignoring the physical limitations of the transmission system, thus producing an artificially 'fair' market in which the generating companies could compete on initially equal terms, but those limitations had to be brought into account at a later stage.

Every day each generating company declares the expected availability and state of readiness of each of its generating sets the next day, and tenders the prices at which it will be willing to keep the set standing by, start it up and supply electricity from it at different levels of output during each half-hour of the following day. NGC then publishes a schedule of intended operation of the sets during the next day, based on that information and on its demand forecasts and the constraints mentioned in Chapter Six (page 201). That schedule is followed except for unforeseen variations in the demand, operating difficulties and breakdowns, etc. The payments to the generating companies are calculated from their declarations and their tendered prices, but the calculations also depend on the network limitations.

Those limitations are that in some parts of the Grid the scheduling is constrained in as much as local stations have to be kept in service because the transmission capacity is too small to enable the local demand to be imported from more economical but distant stations without an unacceptable risk of failure due to a network fault. Consequently those local stations are subject to less competition than the other stations, sometimes almost no competition. The owners of those stations can quote higher prices than they need to cover their costs and make a profit, without any fear that the generating sets in those stations will be omitted from the schedule.

In other parts of the Grid, where the generating capacity is more than enough to meet the local demand but the transmission capacity is too little to enable the surplus to be exported to the parts where the capacity is insufficient, the generating sets in some stations have sometimes to be 'constrained-off', that is instructed not to generate although they could supply electricity at less than the competitive set price in the artificially fair market. In those cases, the generating companies are compensated for their loss of profit.

The volume of information is no greater than it was before privatisation, but incorporating it all into commercial contracts necessitates putting a commercial value on every detail. The calculations involve some 300 equations[13] but the penalties for differences between the scheduled and the actual operations and the compensation for loss of profit to the generating companies by the transmission constraints, though based on physical measurements, cannot be based on the internal station accounts because they are the property of the owners, who are interested parties.

More importantly, the generating companies developed bidding strategies, just as operators in the stock markets develop buying and selling strategies. It is often asserted that such strategies tend to undermine the stability of prices on the stock exchanges, especially when they are triggered by sophisticated computer programmes. The generating companies would be foolish not to use computers to work out their strategies, which juggle with their internal information, their own and their competitors' prices and the rates of penalties and compensation arising from departures from the intended operating schedule.

All the generating companies except Magnox Electric are now responsible to their private shareholders and must be expected to exploit those gaps between the rates of penalties and compensation and their internal costs. They can also use game theory to increase their profits, with NGC and the other generating companies, including the new entrants since vesting day, as their 'opponents'.

Those consequences of the Act and the Pooling and Settlement Arrangements came out in the consumers' complaints and other evidence to the House of Commons Energy Committee in 1991[14]. Soon after that committee had reported, OFFER reported that:

> National Power and PowerGen wished to secure higher Pool prices. The extent to which each company was able to achieve this depended upon the other company adopting a similar policy, in a situation in which the prospect of other generators [i.e. generating companies] bidding to nullify this policy was negligible. I [the director general] do not suggest that the two companies acted in collusion. Nevertheless, I conclude that National Power and PowerGen together have market power, and exercised it in a significant way.
>
> Recent experience suggests that the existence of a duopoly adds to the unpredictability of Pool prices. It is therefore important to ensure that increased competition

in generation provides adequate protection for customers and other market participants.[15]

OFFER had then and still has now no power to regulate the generating companies' prices directly but can refer them to the Monopolies and Mergers Commission. In 1994 OFFER came to agreements with National Power and PowerGen whereby they undertook to divest themselves of four million and two million kilowatts of capacity respectively – and they did so to Eastern Electric in 1996, thus reducing their proportions of the total capacity to the figures in the foregoing table. But neither those interventions nor OFFER's subsequent best efforts have satisfied the professional or general public that the necessary increased competition and better protection for customers had been achieved.[16]

In June 1998, in a meeting at the Institution of Electrical Engineers, representatives of OFFER, the Electricity Pool and the NGC made proposals to increase competition between the generating companies so as to make it more effective[17]. They proposed three markets to replace the existing single market: a forward market for firm prices up to 24 hours ahead of generation, a short-term bilateral market for prices up to four hours ahead and a balancing market for those four hours – to be followed by a settlement process which remained to be defined. In the discussion, a NGC representative opined that it would be essential for the short-term bilateral market to work; 'otherwise it would be incredibly difficult to balance supply and demand' (his words). Other technologists at the meeting were evidently doubtful whether the new arrangements would work any better than the old ones, if at all.

Meanwhile, the new Labour government had published a consultation paper on regulating the privatised utilities[18]. It proposed 'stronger checks and balances within the regulating offices and a stronger element of accountability' by requiring the regulators to consult statutory advisory groups or executive boards or be replaced by commissions, all appointed by the Secretary of State – in effect to restore a degree of political control which the privatisations had been designed to remove.

Economy-sharing regulation of the electricity generation oligopoly

The source of the inadequacies in the Pooling and Settlement Arrangements for the electricity generating companies is in the general assumption – one might almost say the mental fixation – that they must somehow be made to compete with one another. *The new approach in this section is to discard that assumption in favour of allowing them to compete or co-operate, as they may choose, while preventing them from exploiting their customers' dependence on their services to increase their profits at their customers' expense.*

Once the preoccupation with competition is relinquished, the solution can be perceived. It is to apply economy-sharing regulation, as described in the previous section, to each major company in the oligopoly separately, with the same parameters for each. It is a straightforward solution to what has otherwise proved to be an immensely complicated and intractable problem.

The essence of the solution is in the choice it affords to the constituent companies to co-operate or compete as and when they individually choose. One can speculate about how much and in what circumstances they will do either, but it will not matter whether one has guessed well or badly because the companies' best advantage will coincide with their customers' best advantage – as far as either can be calculated – whichever they do and whenever they do it.

The most obvious advantage of co-operation is that it would enable them to exchange information about costs as freely as the component parts of the old CEGB exchanged it before privatisation. They would thereby minimise their daily, monthly and annual costs of generation, the economies being then shared between them and their customers. They could negotiate with one another and with NGC for the most economical arrangements to meet the requirements for safety and security of supply, as laid down by legislation and the regulator. Penalties for non-compliance with co-ordinating instructions could be negotiated with NGC, on which they would inevitably continue to depend for interconnection and bulk transmission.

Such co-operation would not preclude competition, or a mixture of competition and co-operation in other fields. The future course of technological progress is never certain and large organisations often encourage their departments to make independent judgements about the future and back them with investment. The (British) General Electric Company under Lord Weinstock was a notable example, but it is difficult for a single organisation to be flexible in that way – much easier for separate companies to go their separate ways as long as it suits them, but to come together as soon as they perceive an advantage in doing so.

For example, the principle generating companies could take differing views on future combustion cycles, superconductors, optimum size of generating sets and forced draught versus natural draught cooling towers more easily as separate companies than if they were part of a single organisation. In the ordinary course of events, one view would eventually prove to be more successful than the others in each of those four fields. When that became evident, the company which had backed the best development could negotiate licence agreements with the others with no anxiety that their co-operation might be regarded as unlawful or against the interests of their customers. Those interests would be guarded by the over-arching economy-sharing regulation applying equally to them all.

Other oligopolies

Economy-sharing regulation could be applied with equal facility to all the fully commercial (i.e. non-subsidised) public utilities. Its only obvious limitation is that it might be impracticable to apply it in industries which produce great varieties of goods or services, such as the pharmaceutical industry – if indeed that is an oligopoly. The difficulty would arise in computing the customers' benefit from price reductions at different rates over a large range of products.

It remains to judge quantitatively what constitutes an oligopoly – that is to say what proportion of the total turnover must accrue to how few companies for them to be classed as an oligopoly. At present in Britain such judgements are made by the Monopolies and Mergers Commission, but economy-sharing regulation introduces an important difference in the effects of such judgements on the companies concerned.

At present, if proposed mergers or co-operative arrangements between companies are judged to give the parties too much market power, their activities are disallowed or restricted, to their disadvantage in the assumed interests of their customers. But the consequences of economy-sharing regulation are not all to the disadvantage of the affected companies. Permission to co-operate to any degree without external supervision or restriction would be very valuable. So much so that the parties might well cease to assume that they must always resist being classed as an oligopoly in the supposed interests of their shareholders.

Chapter Six concluded with the remark that the science and art of framing tariffs for public services are as yet in their infancy, largely because the relevance of their technologies has not been fully appreciated. The corresponding conclusion to this chapter is that the science and art of regulating monopolies and oligopolies in private ownership are as yet barely born. It seems to be assumed, at least in Britain, that monopolies and oligopolies in the private sector cannot be effectively regulated and that the only remedy is competition – any degree of competition, however puny or restricted, being better than none.

That state of knowledge has something in common with the state of medical knowledge 200 years ago when blood-letting was prescribed for many ailments and caused the premature deaths of thousands of unfortunate sufferers, until medical practitioners began to understand the anatomy of their patients. Unlike blood-letting, which is never or almost never efficacious, competition does promote efficiency in many circumstances, but it is not a panacea, any more than public ownership. Some implications of these conclusions are discussed in Chapter Eight.

REFERENCES AND FOOTNOTES

1) In Britain, franchises have been granted for networks providing telephone and other communication facilities in conjunction with cable TV to households and business premises, alongside existing telephone connections. The legislation did not allow British Telecommunications to instal such networks, although they could have done so at no greater cost.
2) As summarised, for example, in 'The regulation of British Telecom's prices', by the Director General of Telecommunications, July 1988, and discussed in 'Regulating the BT giant, "Consultation" without information', by the author in *Telecommunications Policy*, December 1988.
3) A more descriptive term would be 'profit rate' because it represents the profit over and above the necessary revenue to secure the return – in the sense of recovery – of the invested capital. However, 'rate of return' is adopted in this chapter in spite of that criticism because it is entrenched.
4) However, the 'RPI' in this formula is not the Retail Prices Index itself but the percentage annual rate of increase of the Index – another piece of confusing terminology which has become entrenched.
5) For example the 'formula' given in OFFER's first consultation paper for their Transmission Price Control Review in November 1995 occupied a whole page of equations, including, *inter alia,* a weighting factor related to previous maximum average cold spell demands.
6) First published in the July 1992 issue of *The Political Quarterly* and later, in revised form, in the October 1995 issue of the *Power Engineering Journal*, published by the Institution of Electrical Engineers. The latter article also reviewed a regulation method, known as 'Sliding Scale Regulation', which P. Burns, Professor R. Turvey and Dr T.G. Weyman-Jones proposed in 1995, as a revival of a method applied at the end of the nineteenth century to the then small gas and electricity utilities. The review, which had those authors' consent, concluded that sliding scale regulation lacked three important features, which economy-sharing regulation can provide, but interested readers should refer to their Discussion Paper 11 under that title and their Technical Paper 3 on 'General properties of sliding scale regulation', published by the Centre for the Study of Regulated Industries in May 1995.

7) Two commonly used meanings of 'formula' are 'a general expression for solving problems' and 'a statement of joint aims or principles worked out by diplomats of divergent interests'. The former meaning is intended here, as it was in Chapters Three and Four. However the interests of monopolies do diverge from those of the societies in which they operate and, although the formula has not been worked out by diplomats, it might be accepted by both sides, after negotiations, as a formula in the latter sense.
8) Equivalent to a monetary rate of 8.15% if the inflation rate is 5%, or 13.3% if it is 10%.
9) 'The Transmission Price Control Review of the National Grid Company': Consultation paper November 1995, Second consultation March 1996, Third consultation May 1996, Fourth consultation August 1996, Proposals October 1996 (para. 6.4).
10) The white paper, 'Privatising Electricity', February 1988, Cmd 322, which preceded the bill, asserted that there was 'scope for competition by comparison' in distribution, but the concept was nebulous and was not pursued.
11) One gigawatt is one million kilowatts.
12) As reported in the *IEE Review*, September 1991, by Roger Dettmer, the Technical Editor. In 'The UK electricity pool – a leap in the dark?', he recorded an interview with the Department of Energy (DEn) chairman of those negotiations, in which she said: 'The role of the DEn was very much a facilitator, because a lot of the issues were frankly beyond the competence of officials...'. Nevertheless she became the pool chief executive when she had brought the negotiations to their conclusion.
13) According to an estimate at the Electricity Supply UK Conference at the Institution of Electrical Engineers, 3–4 July 1990.
14) Second Report 'Consequences of Electricity Privatisation', HMSO, February 1992.
15) 'Review of Pool Prices December 1992', Office of Electricity Regulation.
16) The April 1997 issue of the *Power Engineering Journal* carried a Special Feature on the Reorganisation of the Electricity Supply Industry (ESI). Four independent contributors made fourteen serious criticisms of the results of privatisation and of the Pooling and Settlement Arrangements in their articles in that issue, namely:

    Mike Parker, Honorary Fellow at the Science Policy Research Unit, University of Sussex and a member of the Government's

Energy Advisory Panel, in 'The privatisation of the electricity industry: a case of unfinished business':

1. Most of the profits [widely seen as excessive] had been made from the monopoly activities of distribution and transmission and from the coal contracts brokered by the Government;
2. Political intervention was unlikely to be avoided if prices were thought to be excessive or security of supply was thought to be at risk;
3. A fully competitive ESI might become dominated by the financial markets with their emphasis on high rates of return and short-term cash flow.

Dr Nigel Burton, F.I.E.E., Executive Director, SBC Warburg, in 'A city view of the UK electricity industry':

4. The difficulties in trying to reconcile the complexities of the generation market seemed to defy solution;
5. The Regulator's distribution price proposals generally thought to be agreed in 1994 came unexpectedly under review again in less than one year, largely as a result of political disquiet about the level of the REC's profits.

Professor Ralph Turvey, chairman of the Centre for the Study of Regulated Industries and a Visiting Professor at the London School of Economics, in 'Regulation and electricity prices'.

6. The distribution and transmission pricing reviews were enormously complicated exercises;
7. The items in the fixing of X in the RPI-X formula were unavoidably arbitrary;
8. A REC's defence against a takeover yielded better information than OFFER could obtain only a few months earlier;
9. The rival methods of estimating the rate of return required by the capital markets all necessitated intelligent guesswork.

John A. Cazzaza, Chairman of the Board of CSA Energy Consultants, Inc. and President of the American Education Institute, in 'An American's view of the reorganisation of the ESI:

10. The days of engineers sharing information were gone;
11. The cost of money increased because of the larger returns by private investors taking greater risks;
12. The hedging contracts against excessive volatility of pool prices had added to the cost of electricity a component, estimated at between 1% and 3%, which did not exist before privatisation;

13. The stated role of regulation was to prevent prices from becoming excessive and control competitive abuse, rather than to produce economic benefits by reducing costs;
14. Playing the competitive game yielded greater rewards than lowering the costs of producting and transmitting electricity.

17) Discussion meeting at the Institution of Electrical Engineers on 'Developments in the Electricity Supply Industry in the United Kingdom', 23 June 1998.
18) 'A Fair Deal for Consumers', Department of Trade and Industry, March 1998

APPENDIX – FORMULAE AND CALCULATIONS

If b is the base rate,
Q_n is the equity capital at the start of Year n,
s is the sharing factor,
B_n is the customers' initial benefit in Year n,
c is the continuance proportion,
R_n is the permitted additional profit in Year n,
P_n is the permitted profit in Year n,
A_n is the actual profit in Year n,
D_n is the profit deficit in Year n
p is the penalty rate applicable to profit surpluses (negative values of D_n)

and the suffix, $n\text{-}1$, denotes the year before Year n,

then $R_n = s.B_n + c.R_{n-1}$
and $P_n = b.Q_n + R_n + D_{n-1}$.

If $A_n <= P_n$, then $D_n = P_n - A_n$.
If $A_n > P_n$, then $D_n = (P_n - A_n).(1 + p)$.

In the following calculations, the values are stated to the nearest million to show the arithmetic. In the table on page 245, they have been rounded to the nearest ten million to avoid a spurious impression of greater accuracy.

In the first year, the supplier's permitted profit is at 3.0% of its equity capital, £10,000 million, i.e. £300 million, and after a long period of no expansion and no economies, the actual profit is at that level. The customers have had no price reductions, so the total customers' benefit is zero.

In the second year a price reduction of 4.8% overall yields a customers' initial benefit of £203 million and so a permitted additional profit of 3.0 × £203 million = £608 million. There was no permitted additional profit nor any profit deficit in the first year, so the permitted profit in the second year is:

	£ million
Base profit @ 3.0% of £10,500 million	315
Permitted additional profit 3.0 × £203 million	608
Permitted profit	923

but the price reduction reduces the actual profit to £889 million, so the deficit, £34 million, is carried forward to the third year.

In the third year, a price reduction of 4.8% yields a customers' initial benefit of £200 million and the calculations are:

	£ million	£ million
Base profit @ 3.0% of £1,102.5 million		331
Permitted additional profit:		
3.0 × £200 million =	600	
+ 75% of £608 million =	456	
		1,056
Deficit brought forward from second year		34
Permitted profit		1,420
Actual profit		1,387
Deficit carried forward to fourth year		34

And the customers' total benefit in the third year can be calculated as follows:

	£ million
Continuing benefit from second year, increased by 5% due to greater use of the service 1.05 × £203 m	213
Initial benefit from third-year price reduction	200
Total benefit	413

8

TECHNOLOGY AND POLITICS

Politics is the art of the possible[1]

If you are a politician, you will probably have opened this book at this chapter, perhaps wondering whether to take it out of the library or even buy it. The same tendency is to be expected of accountants, commercial managers, philosophers and so on – to start at their respective chapters to see whether they contain any ideas which they may expect to assess easily. But the tendency is probably greater among politicians because they have to make frequent, rapid judgements about whether to read, watch or listen to things as they come along, so as not to miss something important but at the same time not be overwhelmed by the vast mass of information and opinion which pours out of the world's printing presses and broadcasting stations every day.

It is a real dilemma and some books are written so that the chapters can be read with almost equal facility in any order, but that has not been possible here. The purpose of this book is to derive from an understanding of modern technology some important features which a society needs in order to prosper from it. It is concerned with the interactions between technology and the other subjects in the Contents List – accounting, commerce, politics, environmental protection and philosophy. So, if some statements in this chapter happen to support – or conflict with – the reader's opinions, the reason is not that the author has started out with the same – or the opposite – political leanings as the reader. The claim is that the statements are derived from those interactions, which cannot be avoided.

It is a large claim and if you dismiss it out of hand then you might as well put the book back on the shelf. If, on the other hand, you wonder whether there might be something of interest in such a possibility, the only way of coming to an informed conclusion is to begin at the beginning of the book. There are references throughout this chapter to facts and arguments in previous chapters, but they are intended as reminders, not as substitutes for reading the chapters in their published order.

If you go on with this chapter without first reading the earlier chapters (as many readers probably will in spite of the warnings), you may find yourself nodding in agreement with some passages and shaking your head over others, but in both cases you may miss the point that modern societies cannot choose their political principles with impunity without explicit regard to those interactions. Many societies have tried to do so and failed – sometimes gradually and sometimes disastrously – and it can be confidently predicted that such failures will recur. In general, *it is not possible to instal the features which a society needs in order to prosper from technology without an adequate understanding of technology.* That is probably not what the author(s) of the above quotation had in mind, but it is the theme of this book and it applies to politics as much as to any of the other chapters.

THE FAILURE OF COMMUNISM

To start with a robust example of those interactions, let us consider the failure of communism as a system of government. There are some people who still profess to believe in it but most people agree that it has failed catastrophically. However, they do not all agree on why it failed – what are the elements in it which caused it to fail.

An apparently obvious explanation is that in practice communism is cruel and inhuman; it disregards the rights of individuals and relies on arbitrary decrees, coercion and terror to regulate what everybody must and must not do – and even what they must and must not think. But those are not the origins of communism and its few remaining defenders might claim that true communism is a high-minded ideal which has never been put into practice. Its essence is in the slogan: 'From each according to his abilities, to each according to his needs.' to which both Marx[2] and Lenin[3] subscribed, although they did not claim to have originated it. It may have been the basis of some early Christian societies long before communism.

Are the reasons for its failure that human beings are just too selfish and dishonest and that the power which falls into the hands of the rulers of a communist state inevitably corrupts them? Those might be contributory reasons but there is another, entirely separate reason why communism must always fail in a society which relies on modern technology for its prosperity, even if the members of the society are honest and unselfish and their leaders are benign. None of the communist leaders, nor any appreciable number of their followers came anywhere near understanding it, but it can be derived from the earlier chapters of this book, particularly Chapters Two, Three, Six and Seven.

The technological reason is not to be found by embarking on a historical study of the origins, growth and decline of communism. Probably there have been scores of such studies in various languages and there will be many more – in the scholarly tradition of examining just what happened in the past in full detail and (with benefit of hindsight) how and why it happened like that. Technologists have a different tradition and a different approach to their problems. They are not concerned so much with precisely why something – such as a disaster – occurred just when it did as with whether there was an element which made it inevitable sooner or later and which can be changed to prevent a recurrence.

For example, if a civil aircraft crashes, killing a hundred passengers, the enquiry may reveal several contributory causes – snow on the runway, a heavy, gusting crosswind, the pilot was tired – but also that the design of the plane had a flaw.

Even with the flaw, the plane would not have crashed in the absence of the other elements or some of them; indeed it had not crashed on numerous previous flights. But the designers will modify the other planes of that type if they can in the expectation that the other elements are almost bound to recur in the same or a worse combination sooner or later and must not bring about another disaster. The designers of a military aircraft may take a different view in wartime, but this book is not about technology in war.

So in this section the aim is not to forestall the historians in analysing precisely why communism failed so dramatically in the late 1980s in the USSR, but more gradually in China, nor with the semantic question whether those events were themselves the 'failure' or should be described as the 'consequence of the failure'. It is to show the fatal flaw in the communist ideal, which will always make it unworkable. The flaw is embedded in the term 'command economy'.

Stalin is reputed to have decided in 1930 that the USSR should have a canal to carry freight between Leningrad (now renamed St Petersburg) and the White Sea. So he drew a line across a map, prescribed the width and depth of the canal and commanded its construction at enormous cost in human misery[4], only to find that it was too shallow and had other deficiencies which rendered it almost useless. That was a classic example of how the command economy was operated. Neither he nor any of his staff seriously studied the merits of alternative designs, let alone the merits of alternative modes of conveying the freight or of other ways of meeting the needs which the freight was supposed to serve. They had no methodology with which they could make such comparisons. But suppose those communist leaders had all been kindhearted people, incorruptibly intent only on creating their ideal society. How might they have gone about doing so?

To do them justice, neither Marx nor Lenin claimed that they could create an ideal communist society immediately. They promised it as 'a higher phase of communist society, after ... all the springs of co-operative wealth flow more abundantly' (Marx), when 'people have become accustomed to the fundamental rules of social life and when their labour is so productive that they will work voluntarily according to their ability ... [and] each will take freely of his needs' (Lenin). Meanwhile, during the 'first phase' of the new society, Marx had some elementary accounting in mind, with 'cover for replacement of the means of production used up', for 'expansion of production', a 'reserve or insurance fund' and so on. But his ideas on that subject did not amount to anything approaching disciplined cost accounting, let alone the cost method with its balancing of one item against another to minimise the total cost as described in Chapter Three, and they largely lapsed until they were revived, timidly and too late, by Gorbachev a hundred years later[5]. So the question remains: would the ideal society which the early communist leaders sought to create be capable of using technology to produce prosperity?

'From each according to his abilities'

How could the authorities in an ideal communist society measure their population's abilities and bring them into the design of a canal or, reverting to the technology into which we ventured in Chapter Two, an electricity generating station? We have seen in that chapter how the laws of physics by themselves do not and cannot determine how an application of technology should be designed. There is always a range of options, all of which comply with those laws. So, faced with a thousand men and women, how could one measure their abilities to mine copper ore, smelt it, turn the metal into electrical conductors and perform all the other thousands of functions in the construction and operation of the station? How, for that matter, could voluntary workers in such a society measure their own abilities, which are not just their willingness to work? And what relevance would those measurements have to the designers' choices of the materials and sizes of the conductors, the number of stages of boiler feedwater heating and so on?

The answer is, ironically, that in a market economy the testing of abilities has an indirect, beneficial relevance to the design of applications of technology, and the science of testing the specific abilities and interests of school pupils, for example, has made great strides in the last 20 years or so[6]. The direct benefit is that, when the pupils leave school, they can make better career choices, based on objective measurements of what they can individually do easily and well, what they might learn to do with some difficulty if well motivated and what they will probably never be able to do competently, however hard and long they try. They can then choose occupations which are likely to suit them from the range of opportunities revealed by those abilities and interests and the current state of the economy. The indirect benefit is that the designers, who may have themselves taken ability and interest tests, can or will be able to draw on the services of a more capable, better motivated and more contented work force than hitherto.

But in market economies, the designers could and did make their decisions objectively and effectively before people's specific abilities could be reliably measured. Ability testing in that sense is a bonus in a society in which people are largely free to choose what they will do. Whether genuine reliable ability testing is possible at all in a command economy, in which most people have little or no such freedom, is doubtful, but even if it is possible the irony is that the designers cannot make much use of the results in peace time because they lack any comprehensive methodology into which they can fit them. In the absence of established costs and prices, they are reduced to looking over their shoulders and copying what their opposite numbers are doing in the market economies.

Technologists from western Europe who went to work in the USSR before it collapsed came back with scores of anecdotes on that subject. One concerned a Swiss firm which fulfilled an order for some generators. Some years later it was summoned to investigate a complaint and sent a team who had not been to the site in connection with the original order. They were brought to some generators which had the firm's name embossed on them but seemed to be slightly different from their drawings. Eventually they discovered that the generators were imperfect copies of the ones their firm had supplied, which were in another building.

Anecdotes like that were seldom published but they became part of the folk lore of technologists when communism was in the ascendant. They may appear to conflict with the fact that the Russians beat the Americans in the race to put a man into space, but that project was not for the purpose of bringing prosperity to the Russian people. It was an international prestige project, probably with military connotations and almost certainly carried out at the expense of the general population's standard of living.

Where prosperity is the aim, there is of course much to be said for keeping in touch with developments abroad, but only if one understands why one's opposite numbers are taking different paths and whether the factors on which they are basing their decisions exist and have the same weight in one's own projects. Blind copying is a recipe for creeping poverty and sometimes for disaster.

'To each according to his needs'

The slogan refers to personal rather than communal needs, such as the police and armed forces, which are provided on the same basis in non-communist as in communist societies. Perhaps needs sound more directly relevant to prosperity than abilities. Indeed the idea of basing the provision of some services on need rather than payment is pretty deeply ingrained in many democratic countries. But what are a person's needs?

There are some basic human needs which are beyond dispute. Nobody can live for more than a few minutes without air or a day or so without water or a few weeks at most without food, and some clothes and shelter are essential for survival in many climates. But 'needs' meant more than those few things, even in Marx's day, and a much longer list is required if it is to be the basis of a prosperous society. Nowadays we hear about our need for ever more sophisticated medical services, for comfortable, reliable public transport, for heated houses with electric light and even television. One might argue about whether television is a necessity of modern life, without which one loses touch with the rest of society, on which one is heavily dependent, or whether one is better served by newspapers and the radio, but it is difficult to avoid the conclusion that if some people had no television, no radio and no newspapers, in fact no news media (all of which rely heavily on technology), they would be classified as 'deprived' by most politicians and their electors in the democratic countries.

Whether or not the news media are regarded as essential to modern living, it must be clear that a modern society which claims to be governed solely or mainly by the principle of attending to its citizens' needs, without regard to their ability or willingness to pay for them, must supply fuel and/or electricity to their homes and to the factories and farms which supply their other needs. Then how much electricity does each citizen need and when should it be available? If they are all to have as much as they like whenever they like at the touch of a switch, their combined demands will swamp any supply system. Rationing is essential, but on what basis?

In the Second World War, food was rationed early on in Britain and clothing soon afterwards. The rations were so much of this and so much of that per person, with variations for special categories. With the shared objective of first survival and then victory, most people accepted the allocations as tolerably fair. After a year or so of war, coal, gas and electricity came to be rationed – but on an entirely different basis. Each household was required to reduce its consumption to no more than 75% of its consumption in the corresponding quarter before the rationing was introduced. No fairer method could be devised.

The population had of course been exhorted to 'save fuel for victory' and so on before the rationing was brought in, so the extravagant households who had ignored the exhortations got larger rations than the more patriotic households in exactly similar dwellings in the same neighbourhood. Fortunately the war ended and the rationing was eventually lifted; it could not have continued indefinitely on such a basis. In Germany, where rationing was generally stricter, coal, gas and electricity (then produced from coal) were not rationed during the war in the Ruhr Gebiet, where coal was mainly mined, although the miners, who were also mining coal for munitions production, had to be specially exempted from military call-up.

In spite of those examples of the practical problems of fuel rationing on both sides in that war, some devotees of command economics might claim that, with better attention to such factors as geographical location, exposure to winds, building method, size of household and so on, fuels and other things could be 'fairly' rationed for household consumption. But that little word 'fairly' conceals an intractable problem.

To take a simple example, suppose a citizen is considering where to live – whether in a town in a sheltered valley or on an exposed hilltop many kilometres away. He may prefer the hilltop for the view and the nearness of beautiful country for walking, but in a market economy he would be well advised to estimate his heating and transport costs. On the hilltop, unless his house is much better insulated than the houses in the valley, he will need much more fuel, else he will be very cold in the winter and may die prematurely from hypothermia. Also his journey to work will be longer and more expensive. However in a command economy, if the foregoing devotees have done their sums well, that extra fuel and transport will be provided by the state on the basis of need, so he may hope to gratify his love of views and country walking with impunity. And, if he lives in a poorly insulated, badly maintained dwelling, with draughts through the doors and windows and damp in the walls, he will need more fuel than his neighbour in a properly insulated, well maintained house, so he or his landlord will have little inducement to keep his home in good repair. So dwellings will fall into disrepair while fuel consumption rises and standards of comfort decline.

Underlying any assessment of need, there is always a conflict between what is needed and what is deserved. The man's need is genuine; if it is not met he – or his old mother who lives with him – may die of hypothermia, but how did that need arise? Did the man create it? Could he have avoided it and, if he could, should he have taken steps to do so? His neighbour in a similarly dilapidated dwelling on his other side, might say that, although she also likes the view, she has to live there in any case because she is running a farm, with cows to be milked every morning. Her old father, who lives with her, might add that the family has farmed there for centuries and they are not going to be dictated to by bureaucrats from the city who have never milked a cow in the freezing dark of a winter's morn.

That circumstance might be held to change the complexion of the case, but the next question is: does the community need that farm if there are too many dairy farms in the area, perhaps because the population has declined? If not, who is to decide which dairy farmers should be persuaded or coerced into changing their occupations? The bureaucrats? In general, what degree of persuasion, indifference to suffering or positive coercion should the bureaucrats apply to people who may (or may not) know a great deal about farming but who appear to be obstinately set on a way of life which contributes little or nothing to the prosperity of the community on which those people depend for their own prosperity?

It does not take much imagination to see how, once the principle of need is established as the universal basis for decisions of those kinds, the citizens' freedom of choice will be eroded until they find themselves living uncomfortably in crowded blocks of flats miles from any view or country walks – unless, by hook or by crook, they can become privileged persons with 'dachas'. Most people are aware, nowadays, of the abuses and corruption which arise in communist societies, but it is seldom spelled out that they are inevitable in any society which depends on technology for its prosperity but lacks rational costs and prices. That is not to imply that there is no corruption in non-communist societies. There is much more than one would like, but it is not inevitable in the same way.

And the problems are a hundred times as complicated and intractable when it comes to allocating fuel and electricity and other goods and services for industry. Not even the most fanatical devotee of command economics can conceive, even in theory, how fuels could be rationed as between, say, electricity generation and steel production with any real hope of maximising their joint contribution to prosperity in peacetime. The complexity of the task can be gauged by referring back to Chapter Two, which explained (page 39) how the design of a generating station can be greatly simplified at the expense of reducing its efficiency and so increasing its fuel consumption per unit of output. The same is true of steel production but there is no rational way (except the cost method) of deciding how far to tilt the simplicity/efficiency seesaw of either type of plant in one direction or the other.

To take another example, it is no easier to make any rational allocation of electricity supplies between aluminium and steel production without the cost method. Both use large quantities of electricity and either metal can be used for many types of construction. If some technical breakthrough greatly reduced the cost of either metal, it would soon begin to oust the other – and possibly wood and plastics as well – for many purposes in the market economies. But the communist technologists would have no rational way of deciding for which purposes they should make the substitution, when they should make it and to what extent. If they disagreed and made different decisions, no-one would be able to tell in retrospect which decisions had benefited the economy and which had impoverished it. Their only guiding principle would be to try and copy their opposite numbers in the capitalist economies, which they despise.

One could provide any number of such examples, showing that it is quite out of the question to derive prosperity from technology across a whole economy without the cost method. Yet an ingrained mistrust of money leads to its rejection as the basis for many important decisions in Britain and other countries, particularly where safety is involved. We shall return to that subject in a later section.

According to some commentators, the command economy, ruthlessly operated, was effective in the prosecution of the Second World War and it may have saved the USSR from defeat, but that is a separate issue. If the communist slogan had been: 'From each according to his or her abilities and volition, to each according to her or his preferences within what she or he can earn', it might have set the movement on a less catastrophic course, although it probably sounds less high-minded. However, there is much more to deriving prosperity from technology than the ability to avoid a catastrophe, vital though that is. The flaw in the communist ideal was – and still is – a fatal disregard of the cost method in the design of applications of technology in peacetime. Mistakes can still be made with the cost method, but they can be identified as mistakes, and so not endlessly repeated until the very structure of society collapses.

MARKET ECONOMICS

Command economies have failed and will always fail to produce prosperity because their technologists cannot use the cost method to full advantage, if at all, in designing and operating their projects and because those economies do not provide or move towards a system of rational prices, in which the prices of important goods and services reflect their resource costs. Other factors may also contribute to their failure but that one will always apply. Does that mean that free markets offer the only route to prosperity?

The answer is not necessarily because a free market is only one way in which the exchanges of goods and services can be based on their money values. Electricity supply was a state-owned commercial public service in Britain for over forty years, during which time its technologists used the cost method to great effect and its commercial policy struggled towards charging the consumers the resource costs of their supplies. Britain prospered during that period and the electricity supply industry made a large contribution to that prosperity. And the same was true in some other industries and many other countries. In general, a publicly owned commercial public service[7] is better able to contribute to prosperity than a non-commercial public service. The difficult, but not intractable, problem of framing the prices and methods of charging in a commercial public service was tackled in Chapter Six, after some necessary reforms in accounting had been proposed in Chapters Four and Five. It was not necessary to privatise the electricity supply industry in Britain just to enable it to charge the consumers the true resource costs of their supplies.

However some public services are natural or technical monopolies and the question whether they should be in public or private ownership is bound to arise. It is a question on which a non-technologist's opinion may be as good as (though not better than) a technologist's, provided they both understand the issues and do not assume that public ownership in itself prevents the full application of the cost method and commercial principles, or on the other hand that a privately owned monopoly cannot be prevented from exploiting its customers' dependence on its goods or services to increase its profits excessively at their expense without stifling its inducement to economise. The falsity of those assumptions was exposed in Chapters Six and Seven, but unfortunately they remain so deeply ingrained in the minds of the proponents and opponents of privatisation respectively that they invade and polarise many debates about public versus private ownership before the real issue can be brought in.

The real issue – the only real issue if those false assumptions are discarded – is human motivation. Are the people who work in public monopolies or who work and/or hold shares in privately owned monopolies more effectively motivated by the spirit of public service or by self-interest? One's own motives – if one can discern what they are (itself a doubtful proposition) – are of no particular relevance because motives vary from person to person and from time to time. In an emergency, whether short or long, in the author's experience, most people seem willing to work hard, undergo discomfort, take initiatives and suffer hardship voluntarily for a perceived public good if their own and their families' lives are not seriously endangered thereby, and many will voluntarily risk and even sacrifice their lives to save the lives and improve the well-being of others. But monopoly services are seldom concerned with emergencies in peacetime and special arrangements can be made for the few occasions when they are. In the long haul self-interest seems to come out as the consistently more effective motive among people who work in commercial public services (and among those who work in other fields as well, though their motives are not relevant here). Many people hold the same opinion and it is a strong reason for supporting privatisation if – but only if – the legislation includes an effective method of regulating the prices and methods of charging of the privatised monopoly, such as was described in Chapter Seven. The privatisation acts of the 1980s in Britain were seriously deficient in that respect, and they consequently created more controversy and brought less prosperity than they might have done.

Monopolies are at one end of a spectrum. At the other end some sectors of any economy are very concerned with producing or providing a wide range of goods and services. They are generally fragmented and unsuited to be regulated (except for compliance with safety requirements, payment of taxes and so on), even if that was desirable, which it generally is not. Free markets, in which the prices of goods and services are determined by the forces of competition, are generally regarded (outside the remaining communist states and increasingly within them) as the best way of enabling those sectors to make their best contribution to prosperity.

Competition not a panacea

However, enthusiastic advocates of competition often seem to overlook that it is not generally relished by the competitors. They, especially their technologists, are acutely aware of the disadvantages of competition mentioned in Chapter Seven (page 239), which are disadvantages to the whole community not just to the competitors. They would generally prefer to operate on larger scales and with better estimates of the future demands for their products, both of which aspirations are frustrated by the actions of their competitors, and they take no pleasure in responding immediately to external changes – for example in the price or availability of a common raw material – so as not to be outdone or left behind.

Prosperity for them comes just when the competition is weak, either because they have made innovations which give them advantages over their competitors or as a result of favourable (to them) circumstances, which they may or may not have foreseen in advance. They pass on as little as possible of that prosperity to their customers – just enough to attract them away from their competitors, until the competitors catch up or the favourable circumstances come to an end and they have to reduce their prices and/or improve their quality to stay in business.

It is therefore a mistake to expect the suppliers in such a market to co-operate 'responsibly' (the word which is normally used) on some occasions but then to compete 'vigorously' on other occasions. Schemes for self-regulation and co-operative research, much beloved by governments whose officials generally have little or no first-hand experience of competition, are likely to fail or have very limited success for that reason, with the added danger that they also facilitate the exchange of information between the participants, which enables them subsequently to compete less intensively without being found guilty of collusion.

There is no such thing as perfect competition, in which the suppliers have no individual influence on the prices at which they can sell their products. Even competing supermarkets know that most of their customers cannot be bothered to engage in perpetual comparison shopping and that some are incapable of the necessary calculations, so they can discreetly manipulate their prices to take advantage of that inertia and inability, provided they do not go too far. When (inevitably) they do, they have to run campaigns to regain their lost customers and perhaps get a few more. In manufacturing and other parts of the economy where the advantages of large-scale production are greater, the number of competitors must decrease whenever a technological innovation increases the optimum size of the production unit, unless that happens to coincide with an upsurge in the total demand. So the forces of competition will tend to weaken until a recognisable oligopoly is perceived.

Oligopolies resulting from a decrease in the number of competitors in a free market are in a sense inevitable. Oligopolies created by breaking up existing monopolies are in a different category. In some of them, it may be possible to induce the component firms to compete to some extent, but there is no guarantee that the twin disadvantages of losing some of the monopolies' economies of scale and having to manage with inevitably less accurate forecasts of demand, will be offset by that small degree of competition. The current widespread assumption in Britain that even a little competition – or apparent competition – is intrinsically better than a regulated monopoly or oligopoly is a misconception which will inevitably and cumulatively impoverish us until it is relinquished.

When that blind faith in competition is relinquished, the political judgement to be made is the minimum number of competitors below which they constitute an oligopoly and should be regulated as suggested in Chapter Seven (page 258) – not for the purpose of reducing their profits, although that may be one consequence, but to prevent them from manipulating the market prices in the short term and induce them to continue reducing their prices and/or improving their services in the long term.

PHYSICALLY LIMITED RESOURCES

Chapter Six (page 197), referred to the classic economic theory of rent and explained why it is not applicable to the pricing of public electricity supplies derived mainly from thermal generating stations. The reason is that the design and capital cost of a thermal station are not – or should not be – so much affected by physical shortages of resources as to affect the framing of the supply tariffs. Taking the resources – land, labour and capital – in order, the siting of a thermal station is affected by environmental considerations and the quality of the site affects the capital cost (e.g. a marshy site requires deeper foundations), which is reflected in the cost of the supplies but does not affect the number of parameters in the cost models. In a free society, labour and capital can be hired, raised or borrowed in the quantities required for the best design, provided their availabilities are not artificially restricted in ignorance or disregard of the cost method, as described in Chapter Three, or by restrictive practices, which are always inimical to the well-being of those outside the scope of the restrictions.

For such applications, the primary political requirements are a firm insistence on the cost method and rational pricing, coupled with a well-informed determination to guard against price manipulations and restrictive practices. But some applications are also affected by their dependence on physically limited resources. Generating electricity from mountain lakes and rivers and tidal power schemes are examples. We shall consider three resources – land, petroleum (or oil, as it is usually called) and one other, the existence of which is not widely known – to exemplify the political implications.

Land

The prime example of the classic theory (as described in Chapter Six) was land under primitive agriculture, where the fertility of farms depended mainly on their locations. The practice of allowing the private owners to charge high rents for naturally fertile farms, without themselves contributing to their cultivation or bearing any commensurate financial risk, has led to resentment, social unrest, rebellion and war ever since *homo sapiens* began to till the soil. A second factor, which weighs against trusting the theory for non-agricultural uses, is that a plot of land has many features which distinguish it from another plot of the same area and fertility – its shape, elevation, slope, the presence or absence of minerals, its proximity to other plots already committed to incompatible uses far into the future and so on. Thirdly, everyone knows and feels something about land and has views on what should be done with it. The uses to which land is put are part of our concepts of prosperity and the good life, quite apart from economic considerations.

For those reasons, land-use planning has become an immensely complicated subject and no attempt will be made here to formulate any simple theory of it. Suffice to say that the command economy in the USSR ruined huge areas of the country, probably irremediably, whilst the unrestricted play of free-market forces led to appalling urban sprawl and ribbon-development in Britain until they were restricted by the successive planning acts. In the author's opinion, those forces have created hundreds of miles of ugliness down the East coast of Japan and elsewhere. Complicated legislation and prolonged, repetitive, public discussion of every decision seem to be least bad way of planning land use, with the economic theory of rent still in the background but not dominant.

Petroleum

But that does not mean – or ought not to mean – that the theory of rent can play no part in a modern, democratic society. There are other resources whose supplies are physically restricted but which have less complicated physical attributes and fewer different uses. Petroleum is one such. It has a fairly simple chemical composition and only a few uses in its raw state. It is normally processed in a refinery to separate its constituents and modify the composition of some of them. The products can then be used to synthesise a large variety of fertilisers, plastics, medicines and so on or they can be burned.

All those uses of petroleum can contribute to the prosperity of a community but petroleum, unlike land, gives us no aesthetic pleasure. It is not beautiful, it does not smell nice; in fact we prefer to store it out of sight, sound and smell whenever possible and not to think about it, except in the course of our work. So the products of a petroleum refinery can be bought and sold and used without arousing any strong opposition, once the prices of petroleum going into the refineries have been settled. Some uses of petroleum products and some locations of the refineries and petrochemical plants may be opposed on environmental grounds, but those are separate issues which will be addressed in the next chapter.

Whether petroleum is really a physically limited global resource, in the sense that the date of its exhaustion at any particular rate of extraction can be reliably estimated, is a controversial question. North Sea Oil, which is the major source of petroleum under the control of the EC countries, is less controversial in that respect. The total volume of the reserves in the British fields is still not known accurately – or, if it is known the figure is not published – but the period to exhaustion at the present rate of extraction is reckoned in decades rather than centuries.[8]

In those circumstances, successive British governments have adopted the role of landlord in the classic manner, as described in Chapter Three, but for the benefit of the British people rather than that of a few members of an aristocracy, and that policy has provided a rational basis for technologists to apply the cost method to the design of the refineries and the projects which use the refinery products. What remains uncertain is whether the governments which pursued that policy perceived that that rational basis probably contributed more to our prosperity, albeit indirectly, than the revenues which the policy generated. Some people have argued that the revenues weakened our manufacturing industries by raising the international value of the pound sterling and so reducing the competitivity of those industries in the export markets, but they seem to have overlooked the importance of conserving supplies of a physically scarce resource, which is more valuable than it would be if the supply was unlimited.

An abstract resource

There is another, entirely different, physically limited resource which is even simpler in its composition than petroleum, also has no aesthetic appeal and only one, totally utilitarian use, yet the responsible authorities rejected the theory of rent as a basis for its allocation when it became scarce. Their reasons – or lack of reason – are relevant to the main theme of this chapter.

The resource is the radio spectrum. Some readers may not know what it is and some philosophers might be inclined to argue that it has no real existence. It is indeed an abstract concept, though not more abstract than energy, but whatever we may think of it, we all rely on it whenever we watch a TV programme or listen to the radio and often when we make transoceanic telephone calls or calls from mobile handsets. A few words of explanation are therefore in order and are to be found in the appendix to this chapter for the benefit of readers who have not heard of the radio spectrum or would like to refresh their understanding of it.

The use of the radio spectrum is regulated by most governments round the world. They issue licences for transmitters within their borders and co-operate (effectively and for the most part amicably) in their allocations for various purposes, including of course transmissions from ships on the high seas and aeroplanes in the international air spaces. The upper end of the useful spectrum has been raised again and again by radio technologists, but the problems become more severe as the frequencies increase; moreover those technologists have been – in a sense – under pressure all the time from other technologists who have greatly increased the capability and reliability of transmitters and receivers and reduced their size and cost. In so doing they have opened up a variety of new applications and increased the numbers of transmitters in previous applications. Eventually the demand outran the supply in some parts of the spectrum and new applicants could not obtain licences or had to wait an inordinately long time for them. Some action was needed to clear the log jam.

The first instinct of most politicians – and probably of many readers, technologists and non-technologists alike – is to 'leave such matters to the experts'. In this case the experts were some technologists, some economists and some administrators, and the relevance of this example lies in showing what can happen – and very often does happen – when an important problem does not lie wholly within a single field of expertise. For that purpose, some understanding of the problem is necessary.

The first application to be seriously hit by the shortage of spectrum in Britain was civil land mobile radio – mainly radio telephones in road vehicles. The congestion was relieved in

1983 by closing down the old 405-line TV stations and setting up the now well known cellular networks, which use their frequency allocations more efficiently than the simpler networks used by the police and fire services, although at much higher cost. However the independent review[9] which recommended those closures also concluded *inter alia* that the demand for spectrum was accelerating and would in future often exceed the supply under the established procedures for allocating spectrum, which had been developed over many years by the Radio Regulatory Department in (at that time) the Home Office.

The review described those procedures as 'a somewhat arcane business detached from the realities of service and manufacturing industry'. Spectrum had been allocated mainly on the principle of 'first-come-first-served', with minimal charges to cover just the administrative costs; and the internal regulation of each allocation had been delegated to the respective user group. The review found that those user groups – notably broadcasting, defence, civil aviation, police and fire services and radio astronomy – had been allocated 90% of the spectrum covered by the review[10] but that closing down the 405-line TV stations and some other changes would reduce that proportion to 76%, leaving 24% for other civil groups and shared services. The review report remarked that:

> users who organise their own allocations on a delegated basis do not seek spectrum economy for its own sake. They accommodate their own requirements on a non-interference basis but they take their existing allocations for granted and do not appear to review their usage to see if they could manage with less. This is not intended as a criticism. They are behaving rationally since there would be no point in spending time and effort attempting to economise on a resource which is free.

In other words the technologists in those user groups were using the cost method, as described in Chapter Three, to design their projects and (with the probable exception of the radio astronomers) they were logically using more spectrum than they would if they were charged for it. Consequently the latecomers were obliged to invest heavily in super-efficient methods of making the maximum use of their small allocations or forego the advantages (such as time and fuel savings) which radio communication could bring them.

The review committee considered a proposal from the Department of Transport for renting the spectrum and the author later presented it to the Institution of Electrical Engineers[11]. Its specific purpose was to bring the commercial cost of using the spectrum – what the users were charged for it – approximately into line with its resource cost – in this case its opportunity cost in the sense of the cost to the community of denying it to other users. Of course that alignment can never be exact, but the proposal was for the regulating authority to charge a gradually increasing rent wherever the demand exceeded the supply until an approximate balance was achieved. The established users, who would have to pay rent, would then have financial inducements to reduce their demands. They might adopt more advanced technology, share with other users, move to less congested bands, modify their operating procedures or use different means of communication (e.g. land lines), as described in the two references. In so doing they would free some spectrum for other users who could make better use of it and were prepared to pay for it.

The review committee raised no insuperable technical objections to the proposal but stated that its members were not economists and so could not judge its merits. They therefore consulted a group of government economists, who opined that the most thorough-going market approach would be to have a free market in frequencies over the whole spectrum, with users not just renting but also buying and selling rights to use frequencies within a given geographical area. The obvious technical objections to such an unregulated market were that it would lose a set of necessary constraints which the established procedures had developed[12] and would cut across some international agreements. But the author's proposal was to retain those procedures while regulating the demands for licences within those constraints and agreements by gradually increasing the charges to established users and newcomers alike instead of allowing the established users to retain their allocations at the nominal charges, to the effective exclusion of some newcomers.

The administrators in the Radio Regulatory Department, who were neither technologists nor economists, postulated that spectrum renting was intended as a tool for rationing access to the spectrum where the demand exceeded the supply, on their stated assumption that the only way of reconciling supply and demand was to repress demand, and that renting would not promote 'more positive approaches', such as setting equipment standards to minimise interference and encouraging the sharing of frequency bands. But the proposal had precisely the opposite intention, namely to avoid the necessity for rationing by inducing the responsible technologists to make better use of their allocations voluntarily by just such means among others.

In other words, the consequences of 'leaving the problem to the experts' were that the technologists ducked the issue as being outside their province, the economists put forward a proposal which ignored the necessary technical constraints and the administrators presented a travesty of what had been proposed and then opposed it.

Nebulous fears

The two largest user groups, broadcasting and defence, said that spectrum renting would seriously interfere with their operations, but some of their technologists admitted off the record that there was scope for reducing their very large demands without seriously disrupting TV or radio programmes or endangering Britain's defences, as indeed the review committee's comment – that 'user groups who organise their own allocations on a delegated basis do not seek spectrum economy for its own sake' – had made clear. A more widespread and publicised objection, which some readers may instinctively share, was that spectrum renting would somehow deprive the emergency and accident prevention services (fire, police, air traffic control, etc.) of their vitally necessary allocations, with unspecified, catastrophic consequences. Perhaps two examples will assuage those fears.

The first example relates to air traffic control. Airlines have to pay for their aircraft, fuel and air crews and for the airports which they use. An increase in the price of fuel leads to economies elsewhere and/or higher fares and so marginally fewer flights in the short run and possibly to more fuel-efficient aircraft in the long run. It does not lead, even in the short run, to aircraft running out of fuel and crashing. By the same token, an increase in the airport charges, due to the air traffic control services being charged for their frequency assignments, would lead to economies elsewhere and/or higher fares and so marginally fewer flights in the short run and possibly more efficient spectrum usage in the long run. Naturally the airlines would object to spectrum renting but there is no reason to regard those objections as any graver than their dislike of higher fuel prices. The issues are economic, not vital, in both cases[13].

The other example relates to road accidents. In Britain more than 3,000 people are killed and 300,000 injured every year in road accidents, and the suffering would be much greater without our hospitals, doctors, surgeons, nurses, medicaments, surgical equipment, ambulances and crews. They are all vital to saving lives and reducing injuries, yet they are all purchased or hired and no-one outside the communist countries seriously suggests that any other method of procurement would reduce the suffering. Nowadays more lives are saved and injuries reduced by purchasing and installing radio sets in the ambulances and assigning some frequency bands to their communications, but there is no more reason to require those assignments to be free than to require the other links in the chain to be free. Increases in the commercial costs of any of those links must increase the commercial cost of the chain, but those are small prices to pay for the immense benefits of the cost method in reducing the cost and/or improving the service, as explained in previous chapters.

It is even possible that some emergency services would bid for wider bands under spectrum renting than they have been assigned under the old procedures because they could make better use of them (in the way of saving fuel perhaps) than some other users. Such an outcome, which would reduce the cost of the service overall, may be judged unlikely, but trying to guess how technologists would deal with a new situation is a foolish game.

Some opponents of spectrum renting in government departments were probably concerned to protect their jobs, and the author might be suspected of just trying to advance his career. But the emotion which all the opponents appeared to share was that ingrained mistrust of money which has been mentioned several times already in this book. It is felt by many people who have never been concerned with spectrum allocation, North Sea Oil prices or land-use planning, but for whom 'money' is a dirty word. And, at a deeper level in their minds, 'rent' is an even dirtier word, which conjures up images of 'the unacceptable face of capitalism' and rapacious landlords turning out their shivering, rag-clad tenants into the snow.

Some parties in the debate suggested that the word 'pricing' would be less inflammatory in that way than 'renting', and that euphemism has been adopted for another scarce resource which will crop up in a later section of this chapter. But if you do not call a spade a spade, there is always the risk that someone else will do so and attack you for a hypocrite – or that the euphemism will acquire the same bad reputation as the original term. Spades and rents can be used for different purposes. Spades can be used to bury murder victims or to grow vegetables and rents can be used to exploit the dependence of starving peasants on land fertility or to allocate scarce resources to those applications of technology which can make the best use of them, to the advantage of everybody. It is always a mistake to spurn an instrument for a good purpose because it has been used in the past for a bad purpose.

This section up to this point was written in 1994, when it appeared that the cause of spectrum renting had been irretrievably lost. Imagine the author's astonishment when in 1996 the Department of Trade and Industry published a white paper, announcing 'the [then] Government's intention to bring forward legislation to permit the use of spectrum pricing as an aid to effective spectrum management in the 21st century'[14] – on the very lines he had proposed on behalf of the Department of Transport in 1983. That white paper explicitly recognised that the spectrum is a scarce resource, that charges for its use should reflect more closely the value of the spectrum and provide users of congested frequencies with incentives to reduce their demands for those frequencies in all the ways set out in that proposal thirteen years earlier. The white paper eschewed the 'R' word (renting) but it recognised that the temptation to try and maximise the revenue from spectrum 'pricing' must be resisted and it bravely announced that the public sector (including the armed services and the emergency services) should have the same incentives as the private sector and would therefore pay their charges 'on a comparable basis'.

Alas, that euphoria was short-lived. The new government has no intention of charging the large public-sector users although their scope for more efficient use of the spectrum is enormous in comparison with that of the small users in the private sector. It apparently regards the radio spectrum as a legitimate source of revenue from the private-sector users and will probably try to maximise that revenue, thus departing radically from the principle of making the charges reflect the physical scarcity of spectrum. The probable result will be that the principle of using the economic theory of rent to allocate physically scarce resources efficiently will again be discredited in the eyes of the public because it has been wrongly applied.

NON-COMMERCIAL SERVICES

When the British Labour Government was elected in 1945, with a majority of 146 in the House of Commons, its programme contained two opposing strands: to nationalise the means of production and distribution and to provide free medical and surgical services for everybody. They were opposing in the sense that opposite principles were applied to them.

The principles in the nationalisation programme were compulsory acquisition and mainly commercial operation. Thus the country's coal mines, steel mills, electricity generating stations and networks, railways, long-distance lorries and aeroplanes were compulsorily taken away from their owners, who were compensated with fixed-interest government securities, but thereafter the products and services of those assets were sold to the public mainly (though not wholly) on the basis that the customers could have as much as they were able and willing to pay for. There was some deliberate cross-subsidisation, but it was never seriously intended that any of those goods or services should be free to the customers, nor that any of the industries should come to depend on the state for a substantial proportion of their total revenue. Some of them did become dependent in that way, but it was not an article of faith that they should.

On the other hand, although the medical and surgical practitioners were compensated for the loss of their right to sell their practices when they joined the National Health Service (NHS), they were not compelled to join it. However it was an article of faith that there should be no element of payment by the patients for their treatments, either directly or via insurance. Not that the Labour Government was opposed to direct payments or insurance in other fields. Most goods and services (including necessities, like food and clothing) had still to be paid for, insurance of houses and furniture was not discouraged and there was a 15% tax allowance on life assurance premiums under both Labour and Conservative governments until 1985.

The founders of the NHS apparently believed that making all medical and surgical services freely available to everybody would so improve the nation's health that the cost to the taxpayer would gradually diminish. The outcome was that the nation's health did improve, though how much of the improvement was due to the NHS being free was a matter of controversy, but the cost increased continuously and it is now recognised on all sides that the NHS will always absorb as much money as any government can put into it. That realisation has added something to present-day debates about the NHS which did not feature – or hardly featured – in the debates which surrounded its inception.

What the founders did not foresee was the enormous impact that technology would have on the range and quality of medical and surgical treatments in all the industrial countries, irrespective of whether those treatments have been mainly provided by the state, free of charge to the patients, as in Britain, or mainly paid for by the patients, either directly or via insurance, as in the USA, or by a mixture of the two methods, as in France, where the state provides the hospitals and the patients pay the other costs, mainly via insurance. In some respects that technology has reduced the cost of treatment, for example by shortening the patients' stays in hospital – often by a factor of five or more. But the net effect has been to increase the cost, not wastefully but by curing diseases which were previously incurable and restoring to full or partial mobility and health the casualties of accidents and sufferers from systemic disabilities who previously died or were permanently crippled.

Few would doubt that those benefits overall have greatly outweighed their cost. However one consequence has been that the managers, many of whom are of course medically and/or surgically qualified, are faced with unending series of agonising decisions about how to spend the inevitably limited funds at their disposal and how to use the facilities and direct the staff which those funds provide. They have to decide whom to treat and whom to leave untreated, knowing sometimes that they will die; and they have to formulate their own criteria for taking those decisions, since hardly anyone outside the NHS has any pertinent advice to offer and few people are even willing to admit that the situation exists. Meanwhile the technologists – physicists, chemists, biologists, doctors, surgeons, nurses and medical and surgical engineers – keep forging ahead with research and development, sometimes co-operatively and sometimes separately, with no overt constraint except the sums which successive governments and charities allocate to those purposes.

Thus the NHS is a large practical example of the consequences of trying to apply Marx's second principle, 'To each according to his needs', to part of an economy, and it brings us back to the intention, mentioned in Chapter One (page 8), to consider whether applications of technology can contribute to prosperity without being based on money. To address that question, let us return for a moment to the statement at the beginning of Chapter Three (page 51) that the necessary concomitant of the cost method is a rational, practical system of pricing, with the specific purpose of linking the technologists' decisions to the locations at which the benefits of those decisions are expected to accrue. For the designers of an electricity generating station to use the cost method effectively, there must be rational pricing systems for the coal (or other fuel), copper, steel and all the other goods and services which go to the construction and operation of the station. There must also be a rational pricing system for the electricity which the station generates, but for different reasons.

The functions of the electricity pricing system are: (1) to induce the consumers to use the electricity economically, and (2) to stabilise the consumers' demands so that the designers of the supply system can estimate how much electricity to generate and when and where to generate it. If the designers were provided with those estimates from another source, they would not need an electricity pricing system; they could apply the cost method to full effect without worrying about what happens to the energy when it leaves their networks. In fact that is how the junior designers, who fill in the details of the design, work in practice. If they are working on a generating station, they need not even think about the grid. They are told how much electricity the station must generate and when and they get on with the job of generating it as economically as they know how.

Using the cost method in the NHS

So, if the output of the NHS – that is to say the medical and surgical services it provides – could be specified in detail without reference to the patients' ability or willingness to pay for their treatments, the cost method could be used to provide those services economically without prejudice to the principle of a free service to the patients, provided the prices of the goods and services which the NHS requires in order to serve its patients are rational and reflect their true resource costs. Or of course the cost method could be used in conjunction with the patients contributing part of the cost.

That is perhaps how the NHS is or was supposed to operate under the 'internal market' appellation[15], but there is far more to the cost method than switching from direct labour to competitive tendering. To get the full benefits of the method, the managers of the service must understand the details of its six aspects, as outlined in Chapter Three, namely obtaining the estimates, allowing for payments at different times and for inflation, distinguishing between the two kinds of cost, dealing with uncertainties and supervising the execution and completion of contracts. And 'understand' in that context must mean understand both in theory (including the mathematical formulae) and in practice, from experience on both sides of such contracts. The managers must understand what the commercial world means by 'customer goodwill' and 'business morality' and the extent to which they can be relied on, and (preferably at first hand) the temptations which beset those who depend for their livelihood, year after year, on their ability to win contracts by competitive tender.

At present no non-commercial public service comes near to matching those requirements. The most obvious breach is created by allocating budget sums for long-term investments without regard to the operating costs, or (what amounts to much the same thing) requiring the designers of a project (such as a hospital) with an expected life of 50 years to use the payback-period method with a cut-off period of only a few years (sometimes only two or three years). With two important provisos, managers who understand the method will avoid that mistake and experienced managers will allocate sufficient numbers of qualified staff to selecting the contractors, taking account of their previous performance and business morality, and supervising the execution and completion of the contracts. In so doing they will display the customer goodwill on which contractors have to rely if they are to give first class service. The provisos are that the managers' abilities to do those things are included in their job specifications and they are not then frustrated by their own managers, who do not understand and/or are opposed to the cost method. Blind faith in the virtues of competition for its own sake, with no appreciation of its defects, is as harmful as covert opposition. The intention to make the best possible use of the cost method must permeate the whole organisation, up to and through the responsible government minister to the members of parliament who support the legislation on which every non-commercial public service must ultimately depend.

That intention still leaves unanswered the question: what are a person's needs for medical and surgical treatment? The common sense answer, which was the original basis of the NHS, is: as much as medical science and technology can do for him or her. But (as already mentioned) the scope for medical and surgical treatment of an ever-increasing number of conditions is now so large that the total cost of what could physically be done for the patients individually is bound to exceed the total sum which can ever be raised from taxation. So that definition of 'need' leads to the absurd conclusion that our society needs a larger NHS than it can ever have.

That is the challenge to supporters of a free or mainly free NHS – not just to have good intentions and subscribe to high-sounding verbal principles but to put forward a method of allocating the limited resources between the patients, some of whose 'needs', in that common sense meaning, cannot be met for lack of funds[16]. In the continuing absence of any rational response to that challenge, the provision of medical and surgical services in Britain is doomed to remain indefinitely as a controversial subject in the political arena. Many people will be content for it to do so and some will welcome the perpetual controversies, but others may feel that the problem should be faced squarely. It has been caused by technological progress but it cannot be solved by technologists because they cannot bring the cost method to bear on it.

The other non-commercial public services, notably state education, the judiciary, the police force and the armed forces, enter the political arena from time to time for the same underlying reason, but except in wartime, technological progress in those services has not revolutionised the lives and expectations of ordinary people and it is not likely to do so. Thus, although Britain was about a hundred years behind the leading countries in Europe in providing free state education for children, three propositions are now generally accepted (even by those who would like to abolish the NHS): that all children need to be educated, but some parents and guardians are too poor to pay for the education of their children, but our society can afford to meet that need out of general taxation. Educating all our children adequately (in some sense) is agreed to be feasible and an integral part of our concept of prosperity. There is scope for bringing technology to bear on education (more than is generally realised), but there is no expectation that technology will expand the demand for funds for education without limit. Similar considerations apply to the judiciary, the police force and the armed forces in peacetime.

QUASI-COMMERCIAL PUBLIC SERVICES

Some heavily subsidised public services, notably public transport, are in a separate category. They have the appearance of commercial services in as much as they obtain a large proportion of their revenues from their customers, but their dependence on subsidies prevents them from being run on fully commercial principles. Let us label them, for convenience, as 'quasi-commercial services'. Few non-communists would argue that public transport ought to be free for everybody, like education or the NHS, but not very many would defend the proposition that it ought to be fully commercial.

What then are the arguments – including the technological arguments – against subsidising public transport? Although we may have difficulty in defining precisely what we mean by 'need', we all feel that we need the judiciary, the police force, the armed forces (unless we are pacifists) and education and most of us need some medical care during our lives, but can we be said to need public transport in the same way?

When the working population of a city, such as London, expands, the newcomers, looking for somewhere to live, can make simple trade-off calculations. Generally houses of any given size and style will be cheaper the further they are from the places of work near the centre, so the newcomers can balance the higher cost in fares and time of travelling further each working day against the lower cost of housing further out. If they are accustomed to making their own decisions on such questions, they will not all have the same preferences. Some will value their travelling time highly and pay premium prices for accommodation near their workplaces, while others will be content to travel long distances, especially if they can rely on getting a seat on the train to read their newspapers or do some office work. They will also be differently affected by the availability and quality of schools for their children, if they have any of school age. If there are two earners in a household, their choice of where to live will depend on both their workplaces and so on. However most people are influenced to some extent by the cost of their journeys and they will generally choose to live further out if the fares are subsidised than if they are not. If other things are equal, public transport subsidies exacerbate urban sprawl. It is not the only thing which does so – cars are in the same category but we shall come to them later.

If part of a public transport system is under-utilised, an increase in the traffic may increase the revenue more than it increases the cost, but the fares which yield such an improvement are not correctly described as 'subsidised'. If, on the other hand, the costs increase more rapidly than the revenue, a vicious spiral of ever-increasing deficits is set up and persists until the government (central or local) can no longer afford the level of subsidy required to keep the fares from rising. The fares then have to be increased and/or the trains and buses become overcrowded and unreliable, but by that time the subsidised passengers are trapped by their long-term commitments in house purchases, job prospects, children's education and so on.

The situation is drearily familiar to millions of commuters and they come to represent a vested interest in continuing the subsidies at the expense of taxpayers who do not use much public transport. The commuters' need is genuine and they vote accordingly, but it is not the same kind of need as they have for education or medical and surgical treatments. One does not choose to be born or to be ill or have an accident; one does choose where one will work and live. That is the difference.

It may be objected that one's choices of where one will work and live are not perfectly free. Of course not; few if any choices in life are perfectly free; most are made under some degree of external pressure, but they are still choices. It may be further represented that some people have virtually no choice; they cannot move house, they cannot get a job near where they live and if they remain unemployed they cannot meet their financial commitments. But, if their plight is genuine, is it sufficient reason for subsidising public transport for all the passengers? Might it not be less wasteful to give such people concessionary fares or supplement their incomes in other ways, depending on their circumstances?

So, in examining the concept of 'need', we find there are two kinds. Let us call them 'primary' needs, the kind we cannot avoid, and 'secondary' needs, which result from our own choices. One may make wise or foolish choices, based on sound or misleading information, and there may be arguments for holding those (including governments) who give misleading information responsible for the plight of the people who are seriously misled by it, but that is not the same as trying to meet all the secondary needs of the whole population as an object of policy.

That distinction between primary and secondary needs is not totally clear cut. If one takes up the habit of smoking, against the best medical advice, one's subsequent need for treatment for cancer might be classified (though not by the author) as secondary, resulting from one's own choice. That is a question for the NHS managers, though they probably do not use those terms. It is not a question for public transport, in which all the needs are secondary, although some might merit special consideration on compassionate grounds.

Those are the well-known economic and social arguments against subsidising public transport. They are often ignored by advocates of subsidies, although they are probably aware of them. The technological arguments are nearly always overlooked for the more respectable reason that they are largely unknown; in fact it comes as a surprise to most people that there might be such arguments. There are two technological reasons why a quasi-commercial public service will contribute less prosperity to a community than a fully commercial public service.

In Chapters Two and Three, we saw how the designers of an electricity supply system are concerned with the future demand for electricity in the long term – up to 25 years or more ahead of their decisions – and the same would be true of a fully commercial public transport system. Fundamental decisions about the routes of the railway lines and the very expensive tunnels, how large to make the tunnels and the trains, the voltage of the electricity supply and whether to make it AC or DC, the degree of sophistication of the motor controllers (including whether to include regenerative braking) and the signalling system, the balance between buses and trains and many other things would depend on their estimates of the long-term demand for public transport, as measured by the passengers' ability and willingness to pay the fares. For example, the interest on and depreciation of the capital cost of larger tunnels would be balanced against the lower operating cost of running fewer, larger trains or of attracting more passengers by providing larger, less crowded, more comfortable trains.

Those necessary estimates of the long-term demand would of course be difficult to make and subject to errors, but so are the corresponding estimates in commercial public electricity supply. The designers of an unsubsidised service have a definite objective – to meet the commercial demand fully and at the lowest possible cost, with due attention to safety and reliability. On the other hand, if the system is subsidised, those estimates and that objective are undermined. The designers are unlikely to get a guaranteed level of subsidy over the economic life of the capital investments. They may be asked to get the best value for money from a grant for the next few years, but the words 'value for money' beg the vital question: what is the value of public transport to passengers who do not have to pay for it in full? No doubt it has some value, sometimes even a large value, but how can it be measured, and if it cannot be measured how can it be maximised? In truth 'best value for money' is just a catch phrase

The same request has in effect been made for the NHS – here is as much money as the nation can afford, please provide the best value to the patients for that money – which led to the quandary described in the previous sub-section. But at least in the NHS there is general agreement that most people need some medical care during their lives and there is a corresponding commitment to maintaining a level of finance over a long period – not as high or as steady as the managers would like but enough for them to make plans with some hope of implementing them before the money runs out. In public transport, that kind of commitment is notably weaker or absent, which leaves the designers in a worse state of confusion. They may try to conceal it but they cannot avoid it – and it is made worse still if there are capital injections from time to time for extending the system in the vague hope of stimulating economic activity and relieving unemployment in the area. The designers have no methodology for making the best use of such injections.

The other technological argument against subsidies is that they discourage economic developments in other ways of meeting the demands which the quasi-commercial service is intended to meet. In the case of public transport, the affected alternative is telecommunication. Telephones provided a partial alternative to the postal service and fax networks provide a more complete, though more expensive, alternative. It has long been technically feasible for office workers to do a large proportion of their work at home, using video monitors and computers linked by telephone lines to their offices. Or they could work in local offices, in the same building as workers for other employers, instead of actually at home. The facilities are not perfect but they are improving and their cost is falling from year to year. Those improvements would have come sooner if the demand for those facilities had not been undermined by the subsidies of public transport, which became subsidies to workers in city offices at the expense of those who work at or near their homes.

Advocates of public transport subsidies will probably retort that they are essential to relieve the intolerable traffic congestion in cities caused by the much more direct alternative to public transport, namely private cars. They may point out, quite correctly, that one bus, which takes up not much more space than two private cars in the traffic, can replace scores of cars carrying only the driver or, in a few cases, one additional passenger; also that a train can replace hundreds of cars. But traffic congestion is not a logical reason for subsidising public transport. Traffic congestion demonstrates that roads in cities are a physically scarce resource, just as high-density housing shows that land in that area is a physically scarce resource and long, lengthening delays in licensing radio transmitters showed that part of the radio spectrum was a physically scarce resource. The way to share out a physically scarce resource to the best advantage is to charge a rent for its use and to raise the rent gradually until the demand is approximately equal to the supply and the congestion is reduced to an acceptable level.

Road pricing

Road renting – or road pricing as it is usually called because 'rent' is thought to have inflammatory connotations (as mentioned in the previous section) – has been mooted frequently during the last thirty years as the solution to the problem of traffic congestion in urban areas. Its technical and economic feasibility have been demonstrated, for example in Hong Kong[17]. It would be quadruply beneficial in many urban areas in:

1) enabling the public transport system to operate without a subsidy by bringing the commercial costs of its main competitors (private cars) up to their resource costs;
2) discouraging the uneconomic growth of cities, with all their environmental problems;
3) enabling the computer and telecommunication industries to compete with public transport on equal resource-cost terms in bringing about the office revolution; and
4) providing an additional source of government or local government revenue and so reducing the level of taxation or local taxation.

It has been objected that motorists already pay far more than the cost of providing and maintaining the roads via their annual road taxes and their petrol and diesel oil taxes, and that it would be iniquitous to charge them rent for what they have already provided. The answer is that they have paid the commercial costs of the roads, plus a surplus, but not as much as the resource costs because congested roads are evidently a physically scarce resource. It is evidently impossible in Britain, and in many other industrialised countries, to meet the potential demand for motoring and goods transport by simply building more and more roads.

However there would be no overriding objection to using the rents to reduce the motorists' annual road tax, thus reducing the net commercial costs of those who are able and willing to avoid the peak times. It is quite possible that some motorists, for example rural motorists, would be financially better off with urban road pricing than they are at present.

Whether the rents should be used to reduce general taxation or local taxation is a matter for political debate. If local governments had the power to fix the levels of road pricing in their towns, there would be the possible objection that some might be tempted to exploit the dependence of travellers and goods transported through their towns to profit at the expense of that through traffic, in the same manner as some mediaeval towns, which levied duties on goods passing through them and so stifled trade. Indeed it would be iniquitous for local governments to take through traffic, as it were, unawares by giving it not enough time to take other routes, or otherwise adjust to the charges, and that consideration might severely limit the permissible rate of increase of road pricing charges.

Moreover it would be important to prohibit any discrimination in favour of local traffic and against through traffic or vice versa. The charges should be determined by the degree of congestion, not by the origins or destinations of the vehicles. But with those two conditions, there is no technological objection to local governments gradually reducing their urban congestion to the levels voted by the residents. They would of course be wise to balance the loss of local trade which any diversion would cause against the revenue and the benefits of a reduction in congestion.

It has been suggested from time to time, as an alternative to road pricing, that a tax on parking at work places in town and city centres would relieve traffic congestion by persuading commuters to travel to and from their offices by bus or train, or even by cycling or walking, instead of in their cars[18]. Such measures might succeed – and may have succeeded in traffic experiments – for some months. But the suggestion overlooks the probability that, in the long run (several years), the commuter traffic would be replaced by other traffic, such as intending through traffic which is at present held back by the congestion.

For example, the M25 orbital motorway round London was built largely to relieve urban congestion by offering a quicker route to the traffic which was bound for the other side of London. That congestion did diminish for a time, but after a few years that through traffic was replaced by new commuter traffic, so that now the congestion is as bad as ever. The M25 is also now congested at times, so intending through traffic would return to the urban routes as soon as they were found to be less inconvenient.

Traffic engineers knew 30 years ago that congestion was the principal limiting factor on traffic movements within 50 miles of the centre of London, and in corresponding circles round other cities and towns. But that was not understood – and is apparently still not understood – by transport policy makers. So great is the convenience of and potential demand for private transport (passenger and freight) that nothing except physical barriers will relieve the congestion as long as the roads are free at the time of use. But physical restrictions, unless they are total, must depend on systems of permissions for some vehicles to pass, but not others. The classifications in such systems are inevitably cumbersome, arbitrary, expensive to enforce and subject to evasion. They go against the principle of making the most economical use of a scarce resource.

In short, road pricing is not the best of a number of solutions; it is the only rational, practical solution to the problem of traffic congestion on urban roads.

REMUNERATION

In aiming to make pay a reward for ability and effort, the slogan 'A fair day's pay for a fair day's work.' may seem to be an improvement on Marx's ideal, earlier in this chapter. And attempts have been made from time to time to define and measure what is fair, but with no success. Tom Brown got to the kernel of the problem when he remarked 140 years ago: 'He never wants anything but what's right and fair; only when you come to settle what's right and fair, it's everything he wants and nothing that you want.'[19]

So what should be the principle by which people should be remunerated for work in a free society with a monetary economy? Technologists do not have far to look for it. In a well-run society with full employment, labour would be scarce – and therefore valuable. Inevitably some kinds of labour would be scarcer than others and different kinds would be scarcer at different times. The term must be 'kinds' rather than 'grades' of labour because, although nearly everybody would agree that the skill of a brain surgeon is of a higher grade than that of a butcher, it would be impossible to obtain agreement to a list in which all skills were graded. Some people might think that cutting coal in a mine requires more skill than fabricating bled-steam boiler feedwater heaters in a factory, but others would not agree. On the other hand we should all agree (if we could be bothered to think about it) that those jobs require different kinds of skill, in the sense that a skilled coal cutter would not be able to fabricate heaters without retraining and vice versa.

Thoughtful technologists also know that, although some skills are more easily acquired than others, the notion of wholly unskilled, but still useful, labour is another illusion. When it comes to cases, some people can hew wood and draw water more quickly and become less tired than others; moreover beginners can almost always be trained to do those things more quickly and better than they are likely to discover if they are left entirely to their own devices. And the same applies to digging trenches, humping bricks and the other so-called unskilled jobs.

On those bases, the remuneration for any kind of work in a free society should be whatever is required to induce enough people to acquire the skill to do that work and offer their services in the places where that work is required. Then, on the other side of the equation, the number of people with any particular skill which the designers of applications of technology will want to have working in any location will depend on their remuneration – the greater the remuneration the more the designers will look for substitutes in order to minimise the costs and so the smaller the number of people they will want to employ, or recommend their managers to employ. On that principle, the designers' efforts to economise and the workers' preferences for well-paid jobs will combine to persuade approximately the necessary number of people to engage in applications of technology throughout the economy.

Of course that principle was not invented by technologists. It is part of standard economic theory and it is not restricted to applications of technology. It is consonant with the cost method and, like the cost method, it has its warts, but it is not to be dismissed on their account unless and until there is something better to put in its place. A method based on fairness in some sense, even if fairness could be defined, would fail to provide the vital link between the supplies and demands for all the kinds of labour and skills on which a modern industrial economy depends. However some enlargements and some *caveats* are desirable.

The principle does not force people to take up the most highly-paid occupations of which they are capable. They may have as much or as little regard as they choose and can afford to what interests them, satisfies them and/or what they judge (by their own lights) to be most valuable, as well as – or even to the exclusion of – the remuneration. Some people with unearned incomes and/or pensions are willing to work – sometimes quite hard – for no remuneration, but in present conditions most people have regard to the remuneration for working as well as its other attractions and detractions and not nearly enough necessary work would be undertaken voluntarily without payment, even in the most prosperous societies. Also some jobs have to be paid much more highly than others to attract enough people to do them – because they are tiring, dangerous, dirty, lonely or beyond the abilities of most of the population. Seen in that light, remuneration is not a reward for industry *per se* but for doing what is necessary but would otherwise not be performed.

An essential lubricant (all too often absent) for the smooth working of the principle is information. Just as in Chapter Seven (page 240) it was seen that the tenderers for a contract cannot compete effectively if they do not know the prices at which similar contracts have recently been awarded, so people cannot choose their occupation with due regard to remuneration if they cannot find out how much other people are being paid for similar work. Perhaps the most beneficial reform in British employment practice would be for the remuneration of all employees to be published in the annual accounts of their employers (available for a fee to cover no more than the cost of producing the information). But of course such a reform would go against the love of secrecy which permeates our society.

Trade unions and professional associations

In about 1986, the leader of one of the British print unions claimed proudly in a televised speech that the union had successfully resisted 'the new technology' in newspaper printing for sixteen years. A few months later, *The Times* newspaper, in a surprise move, dismissed its print union employees and moved to new premises in Wapping, where that technology, comprising computer type-setting and other features, had been installed. A few days later, the paper carried an article which gave details of the union's restrictive practices before that move. It recounted numerous instances, across the whole production and delivery of the paper, of far more men (few women were involved) being paid than actually did any work, or even turned up at the premises; indeed it would have been physically impossible for them all to be present during the times when they were nominally working. According to the article, those practices, which had been growing in extent throughout the twentieth century, had resulted in three times as many men being employed on a printing press in London as on the same type and size of press in Chicago, San Antonio, New York or Sydney[20].

Many technologists, including the author, could recount at first hand their experiences of restrictive practices by other unions and the damage they inflicted on productivity and delivery times, but they largely have not attempted to publish anything for fear of precipitating strikes. The foregoing article stated explicitly that 'any attempt to let the outside world know what was happening would have led immediately to a strike'.

Some readers may suspect that that article was biased against the unions, but it was largely confirmed in 1992, when Eric Hammond, who had been an active trade unionist throughout his working life and became the General secretary of the Electrical Trade Union, wrote a book about trade unions, their restrictive practices and how they fomented strikes from his own first-hand inside knowledge. In Chapter Six – 'Wapping, point of no return' – he recounted that 40% of Fleet Street printers were over the age of 65, seldom worked more than two days out of five before being rostered and spent the rest of their time in other jobs; hundreds were in their 80s, some could not climb the stairs to collect their wages and sent their relatives to do so; strikes were almost a daily event – and so on.[21]

The power of the unions was drastically reduced in the 1980s and there are now far fewer strikes and restrictive practices, but in as much as trade unions and professional associations, when they are concerned with remuneration, are still able to influence the bargains between employers and employees, there seems to be a *prima facie* case for some form of regulation. However it does not follow that they should be treated in precisely the same way as has been proposed in Chapter Seven for monopoly and oligopoly suppliers of goods and services.

The difference is that the purpose of allowing the suppliers of goods and services to make profits is to harness the self-interest of their shareholders and work forces to co-operate in improving the quality and reducing the prices of what they supply. Society has no duty to guard the welfare of suppliers of goods and services in that capacity, except to treat them justly. On the other hand many people believe that society does owe a duty to guard – to some extent – the welfare of those of its members who are manifestly incapable (for one reason or another) of bargaining effectively with their employers or potential employers for their remuneration and their working conditions.

Seen in that light, the one service which trade unions and professional associations, when they are concerned with remuneration, can claim to provide is bargaining ability on behalf of their weaker members. Better education and regulation of working conditions might be more effective remedies for that weakness, but it may be that some people are constitutionally incapable of bargaining effectively – and moreover that attempts to teach them the art (which inevitably involves bluffing and other deceptions) would seriously weaken their more valuable abilities.

The problem, therefore, is how to prevent the unions and associations from exploiting their power to restrict the supply of labour artificially in the financial interests of their members, while still enabling them to protect their weaker members from unrestrained exploitation by their employers. In most people's eyes, it is a moral problem but it is also important in the context of obtaining prosperity from technology.

UNEMPLOYMENT

An unwelcome feature of technological progress has always been that it changes the demands for different skills (often quite rapidly) and so causes labour redundancies – as the French peasants who threw their sabots into the machinery which was taking away their livelihood discovered and tried to remedy. It is surprising how often that simple proposition is evaded. Sometimes the redundancies can be reduced by rearranging the work, or the affected workers can be retrained for different jobs, but such counter-actions do not happen automatically.

It follows that a society which aspires to prosper from technology must have an education system which gives all its young people when they first seek gainful employment the ability and willingness to do at least one job for which there is a demand and – not less importantly – the ability and opportunity to learn other jobs as and when the demand for the skills which they initially acquired are overtaken by further progress. That may sound rather trite but the British education system is still grossly deficient in that respect, for two political reasons. The first is that Britain was a hundred years behind the leading European countries in introducing any kind of free state education and has still not caught up.

The second reason, related to the first, is that many politicians believed until recently (and some still believe) that training for a specific job is not very important, possibly because politics, as an occupation, is one of the few for which there is no recognised prior training. Or they believed – and still believe – that 'vocational training', as it has come to be callcd, is 'the responsibility of industry', without thinking what they really mean by 'industry'. Industry is yet another abstraction, embracing all sorts of organisations, large and small, some obliged to compete vigorously to survive whilst others are monopolies or members of oligopolies. Keenly competing firms may decide to train their own work force but they will not do so if they estimate that they can get what they need at less cost by recruiting fully-trained people externally; indeed if they did so against their commercial judgment they might not survive. Rather more might be expected from (or imposed on) monopolies, but even they will be reluctant to train people who are unable or unwilling to benefit from it.

It might be thought that the obligation to retrain employees who are made redundant by technological progress should be laid on those who introduce the new technology. The objection is that the redundancies generally occur in parts of the economy remote from the new technology, the classic example being the internal combustion engine in road vehicles, which caused redundancies among railway workers all over the world; and more recently computers have made clerical workers redundant. There is no way of calculating how much disruption any new technology will cause or the cost of retraining the affected workers, and of course there is the additional complication that the new technology often comes, wholly or in part, from another country. Vocational training must therefore be the responsibility of central government, except in so far as it can persuade firms and other industrial organisations to take it on voluntarily, and it is not a responsibility which can be shifted to local government because the innovations are generally outside the areas where the redundancies occur. That responsibility has been willingly shouldered in some countries, but in Britain vocational training is the Cinderella of our education system.

It is important to emphasise that the foregoing advocacy of government involvement in vocational training is derived from technological considerations. Unfortunately political opinions tend to become generalised and polarised. In this case one group is always looking for less government involvement in the organisation of society and the other group always embraces any extension of it. The former group opposed the Labour nationalisation programme and the formation of the NHS; it supported the Conservative privatisation programme; it tries to weaken the NHS whenever it can and it refrains from advocating its abolition only because it senses that that would be electorally disastrous. It opposes, overtly or covertly, central government sponsorship of vocational training. The latter group supported the creation of the NHS and nationalisation; it opposed privatisation and it refrains from advocating renationalisation only because it senses that that would also be electorally disastrous. It supports government sponsorship of vocational training but blindly, which augurs ill for its organisation and cost. Neither group is seeking objective criteria for judging whether government should take on a particular activity or not.

Our examination of those questions has not uncovered any overriding technological arguments in favour of or against private ownership of the means of production and distribution or whether medical and surgical services should be free. However what can be said with confidence is that a society whose central government takes responsibility for the vocational training and retraining of its people will be more prosperous than an otherwise similar society which leaves those services to be shaped by the forces of free markets.

That is not just a plea for more spending on education. For in the words of a recent report:

> No significant statistical relationship has been found between levels of educational expenditure within advanced industrial nations and national educational outcomes. Of the countries in this study, Japan has the lowest public expenditure per student and the highest educational outcomes. The USA has the highest public expenditure per school student and the lowest educa-

> tional outcomes. Britain comes above Japan and France and below Germany and the USA on per student spending and also has an intermediate position on the percentage of GDP spent on education (above Japan and Germany; below France and the USA.).[22]

The essential change in Britain is for its legislators and the voters who elect them to take vocational training seriously for the first time in our history – and then have the humility to admit that we are not in the vanguard and that what we have to do is to stop trying to work out our salvation from first principles, and to copy our more successful European neighbours[23].

Whether the activities of trade unions help to reduce unemployment or make it worse is a matter of further controversy. What can be said is that strikes and restrictive practices are more obviously harmful to the community when there is full employment than when there is substantial unemployment. For under full employment the lost production cannot be made up and the benefit (if any) to the strikers is all at the expense of the rest of the community, in the form of higher prices and/or disrupted supplies and/or services. But full employment is precisely the source of strength which enables the unions to strike and engage in restrictive practices most effectively. So, if trade unions are to be regulated to prevent them from using their strength to distort the bargaining process to the detriment of prosperity in the long term, the regulations should be strictest when there is full employment.

A corollary of the cost method

Better vocational training, while essential as a means of reducing unemployment, will not solve the problem. Chapter One made it clear that this book does not offer a cure for unemployment. However, it happens that the cost method has a corollary relating to unemployment which has not been much debated.

The corollary stems from the difference between resource cost and commercial cost. As explained in Chapter Three (page 88), applications of technology make their best contribution to prosperity when their designs are based on their resource costs, which are what the community has to forego in order to obtain the benefits of those applications, but their designers normally base their designs on the commercial costs, which are what the organisations which make the applications must actually pay. So any large differences between the two kinds of cost are potential threats to prosperity.

An important part of the cost of any application of technology is the cost of the people employed in it – in designing it, manufacturing its components, constructing it, operating it and sometimes decommissioning it and disposing of what remains. The commercial cost of the people is what they are paid, plus the cost of their accommodation at work and the employer's pension and national insurance contributions, but what is their resource cost? How much does the community forego in order to employ them on that project? When there is full employment it is the same as the commercial cost because if they were not employed on that project they would be employed elsewhere in the economy, but when there is substantial unemployment that is no longer the case because the alternative to employing them on that project is to pay them unemployment or welfare benefit in one form or another.

Let us return to Chapter Two (page 32) for a moment and reconsider the question of how many bled-steam boiler feedwater heaters we should instal in our generating station. The effects of installing an additional heater are to save some coal throughout the life of the station at the expense of fabricating the heater, with its associated pipe work, adapting the turbine to allow more steam to be bled from it, enlarging the turbine hall and some other minor changes. In that chapter we could not decide how many heaters to instal just by applying the laws of physics and common sense without the cost method, but now let us suppose that, having employed the cost method correctly, we find there is a severe shortage of fabricators in the heater factory, while the mines find they cannot sell all their coal at the prices which will cover the miners' wages and other costs.

The free-market theory is that, in such circumstances, the miners' wages should be reduced and the fabricators' wages increased, thus lowering the price of coal and raising the price of the heaters, with two sets of consequences. The first is that, seeing those price changes, the designers of the next station will instal fewer heaters and buy more coal, which may redress the balance between the demand for heaters and the demand for coal. If it does not, or it does so only partially, the second set of consequences is that poverty, or the fear of poverty, among the miners, and greater prosperity, or the expectation of greater prosperity, among the fabricators will induce some miners to retrain as fabricators and move from the mine to the heater factory. Those are the mechanisms by which the people in the community are expected to adapt themselves to technological progress, and the only modification proposed in the previous section is that the government should shoulder the burden of retraining redundant workers except to the extent that it can persuade employers voluntarily to undertake it.

The disadvantages of such mechanisms are that in practice they are slow and cumbersome in their operation and lead to miserably long periods of high unemployment in some areas, even in the countries with first-class retraining services. So much so that they are almost never allowed to operate unimpeded in a modern economy. The inducements to adapt to the pressures of technological progress are weakened by the payment of unemployment and welfare benefits, in various forms and under various names, to which the people affected have often contributed in advance by their national insurance payments. Moreover the areas of high unemployment sometimes become so large that it is no longer certain – or even probable – that retraining and moving people would keep pace with technological progress.

Returning to the difference between commercial cost and resource cost, the resource cost of employing people in an area of unemployment is their commercial cost, less the employer's national insurance contributions and what those people receive for remaining out of work. If potential employers in the area could obtain the services of unemployed people for only that cost, they would often employ more people and so increase the prosperity of the area without any increase in the flow of unemployment or welfare benefits into the area. Let us call that kind of employment 'resource-cost' employment.

The aim is simple and straightforward – to lessen the gap between the commercial cost and the resource cost of employing people. The essential features are that, when employers take on people who have been unemployed for substantial periods and would otherwise receive unemployment or welfare benefits in one form or another, those benefits should not cease or be reduced immediately but should be diverted to the employers, who should also not have to pay the full national insurance contribution, as inducements to employ those people; however those inducements should taper off as and when the level of unemployment in the area diminishes. As in the proposal in Chapter Seven for economy-sharing regulation of monopolies (page 242), the details are less important than the principle, but again there is no point in making the rules more complicated than necessary.

The first thing to be decided is the level of unemployment at which the inducements should come into operation. Five per cent might be a suitable level at which to begin reducing an employer's national insurance contributions and diverting the unemployed person's benefits to that employer, with the reduction and diversions increasing from zero below that level to 100% – including all the benefits, allowances, etc. which the new employee no longer receives and the employer's national insurance contributions – when the level reaches or exceeds, say, 15%. To obviate anomalies resulting from different levels of unemployment in adjacent areas, the level applied to a particular point of employment could be weighted by its distance from the nearest area boundary, so that the level applicable at the boundary would be the average of the levels in the two areas and the applicable level at the centre of an area would be the level in that area.

It is of course long-term unemployment which causes the really severe distress and demoralisation. The arrangements need not and must not cater for temporary unemployment, which is unavoidable in any market economy. There would therefore have to be a waiting period for the unemployed person, probably at least six months (with normal unemployment payments during that time), before any payments to a new employer became due, again starting from zero and rising to the maximum after, say, one year without employment. Thereafter the payments would continue at that commencing rate for, say, a further year before falling to zero over the year after that. To cater for temporary employment, the waiting period could be waived after the first period of unemployment until the employee had been employed for a year, reckoning periods of temporary employment cumulatively. And it would probably be desirable to commence then with a shorter waiting period and a faster build-up of payments until the cumulative period of employment had increased to, say, three or four years.

There is no point in setting down any precise formula because its shape and its parameters would have to depend on factors which cannot be assessed quantitatively *in vacuo*, but that does not mean they should be left to the discretion of government ministers or officials from time to time. The rules should be simple and published, so that they can be studied by potential employers and the designers of projects and included in their calculations, with confidence that they will not be changed at short notice. Potential employers would have to make their own forecasts of the future level of unemployment, or they might rely – at their own risk – on consultants' forecasts, but they should be able to calculate with confidence how much they would receive in respect of any known group of employees at any forecast unemployment level. There could be parallel arrangements for self-employed people and for part-time employment.

Some opponents of resource-cost employment might try to condemn it as a variant of 'workfare', whereby the receipt of welfare benefits depends on the recipients doing work which they may feel is demeaning. But – without prejudice to any judgements either way about workfare – that would be a false description. To the employees, resource-cost jobs would have the same outward appearance as any other jobs. Some employees might not even be aware that the jobs were subsidised and in any case millions of people have worked for most of their lives in subsidised industries without feeling in any way demeaned thereby.

It might nevertheless be more legitimately objected that the fall in the level of unemployment might be intermittent, with people being employed in short-term projects which would end when the payments to the employers dried up. But for most people a temporary job is better than no job when there is widespread, long-term unemployment in an area. And there would always be the hope that some of the new jobs would be or become more permanent.

A variant of that objection is that employers might dismiss unsubsidised employees and replace them with recruits from the pool of long-term unemployed, and then dismiss those recruits in turn when their subsidies ran out. Theoretically resource-cost employment might do little more than rotate the available jobs in that way and so take up the subsidies without appreciably reducing the level of unemployment overall. That would be an inefficient way of running a business but, if the worst came to the worst in that way, the question would be whether it is better for morale for some people to have secure jobs while others have no hope of a job, or for the available jobs to be offered in rotation to all or most of the long-term unemployed. And if (contrary to the author's preference) the former outcome is preferred, the next question would be: is that preference strong enough to override the hope that some of the new jobs would be genuinely new and long-lasting?

Another theoretical scenario is that the subsidies paid to employees in an area of high unemployment might enable them to undercut their competitors in an area of less unemployment, thus reducing unemployment in the former area at the expense of the latter. In other words the resource-cost jobs would show a tendency – though not necessarily a strong tendency – to even out the level of unemployment between different parts of the country. If that happened, it would reinforce the policies of most governments in injecting financial aid into high-unemployment areas at the expense of the taxpayers in the areas of low unemployment.

Resource-cost employment is not put forward as a cure for unemployment but as a mitigant – to lessen its worst effects. Two of its political advantages are that (unlike many of the proposals in this book) it is not fundamental and (unlike most proposals for reducing unemployment) it would not make any additional drain on the public purse. It might be criticised as requiring a continuing inflow of public funds which would not be necessary if the level of unemployment could be reduced by other means. But of course if such other means were available they would take precedence automatically, since resource-cost employment would not come into operation until the level of unemployment was already unacceptably high. Seen in that light, resource-cost employment would be a way of using unemployment and welfare benefits to better advantage – making them contribute to prosperity by creating jobs which would otherwise not be available.

However resource-cost employment, combined with first-class vocational retraining, would still not get to all the roots of the unemployment problem. For technological progress does not just change the relative demands for different skills; it also reduces the demand for any kind of skill – indeed any kind of labour to achieve a particular level of prosperity in particular circumstances. Consequently a society which allows and encourages technological progress must not only have the right kind of educational system; it must also aspire to an ever-increasing level of prosperity, or suffer from ever-increasing unemployment, unless its circumstances change.

Large wars are the classic cures for unemployment, but in the long run it is also reduced by gradual changes in the current conception of prosperity and by non-military threats to existing prosperity. The gradual changes comprise demands for new goods and services, not previously available, and for much more leisure than most people could hope for, even a few decades ago. The non-military threats to existing prosperity are an aging population and damage to the environment (which is tackled in the next chapter). Both must be expected to pose difficult problems, but at least they will also help to reduce unemployment, unless they are very badly handled. Unemployment is, in a real sense, caused partly by prosperity, so that reductions in prosperity ought to result in less unemployment.

PRACTICAL POLITICS

A friendly critic of this book, who was persuaded to comment before reading this chapter, wondered how it would deal with decision makers who, when all the technological, accounting, economic and other 'objective' aspects of a subject have been put before them, apparently disregard them for the sake of quite different considerations: perceptions of electoral advantage, self-interest (keeping their jobs, or even 'not in my backyard'), popularity, prestige, 'face', previous commitment, or even sheer obstinacy or prejudice.

The answer is that it is impossible to prevent politicians – or indeed non-politicians – from disregarding logical arguments and long-term consequences in any kind of society. They rely in the end on their own judgement, which is sometimes wise, sometimes foolish and sometimes even wicked, but they have much less difficulty in hearing or reading about those arguments and consequences and much more difficulty in disregarding them in a democratic society than in a dictatorship or totalitarian regime. That is of course one of the reasons for preferring democracy. However, in a democracy, the politicians in power have been selected automatically for their desire and ability to get themselves elected. So they are generally unwilling to promote legislation or espouse policies which they believe will lose them too many votes at the next election. Under strong leadership they sometimes persist with policies espoused by the leader which most of them mistrust, and sometimes the leader is vindicated by a subsequent change of public opinion, but in the continuing absence of such a change they are bound to bow to the wishes of their electorate in the long run – or be replaced.

That does not mean that everybody must be enthusiastic about – or even fully understand – any particular measure before it can be implemented, but the tacit assent of the electorate is an essential ingredient. The most that any author can do is to clarify the arguments, reinforce some and, where necessary, radically revise others.

So reviewing the six main sections of this chapter *seriatim* in that light, communism is now widely discredited, but not for the reasons set down in that section. *The Times* columnist, Bernard Levin, foretold the demise of communism in the USSR and its satellites – with uncanny accuracy – in 1967, after the Prague Spring, that is to say 22 years before it occurred. According to him, even the communist leaders were beginning to realise by then that, as a system of government, communism failed to bring home the bacon – or in his idiom the potatoes. But for the most part its collapse seems to be accepted with little curiosity as to the root cause and no understanding that it had anything to do with the nature of technological decisions. There is therefore a real danger of its revival, which can only be prevented by exposing its lack of – indeed denial of – the essential feature without which no society can prosper from technology, whatever other features it may have.

The section on market economics will have largely succeeded if it discredits three facile assumptions: that the main choice in economic policy is between state ownership and private ownership of the means of production and distribution, that competition is the only effective way of inducing commercial organisations in the private sector to economise and pass on the benefits to their customers and that even a little competition is better than none. In truth there can be every degree of competition from almost total monopoly to almost perfect competition in any prosperous non-communist economy. Monopolies not in the public sector and oligopolies can be regulated as such much more effectively than they have been in the past, without undermining their inducement to economise. Consequently the most important political judgements to be made are the minimum number of competitors below which they constitute a monopoly or an oligopoly and should be, not broken up but publicly regulated.

The thrusts of the third section are: (1) to revive the economic theory of rent as the only rational way of allocating the consumption or utilisation of physically scarce resources to the best advantage, without the unwanted social consequences which have come to be associated with that theory, and (2) to show what can happen when important problems are 'left to the experts' without making sure that they do not overlap the boundaries between two or more of the established fields of expertise.

The fourth section puts forward and defends – on technological as well as social economic grounds – the proposition that quasi-commercial public services in general and public transport in particular should become fully commercial. Unfortunately it has come to be almost universally assumed that subsidised public transport offers the only solution to ever-increasing traffic congestion in towns, so the proposal is not likely to be seriously debated in the near future. Public assumptions do change over time, however, as witness the attitude to drink-driving, which has changed so much over the last 30 years or so that it is difficult to recall that driving while inebriated was once accepted as tolerable, not reprehensible and even slightly amusing. So perhaps the assumption that public transport must be subsidised will eventually be called in question.

Remuneration for work done, the subject of the fifth section, has always been a vexed question in civilised societies. The purpose of prospering from technology is only served to the extent to which remuneration is accepted, not as a reward for working hard or for obtaining particular qualifications *per se* but for acquiring the ability and necessary qualifications and then doing work which would otherwise not be done, or not done properly. That principle also goes against public opinion that earnings should be fair in some sense, especially when it is reinforced by political speeches crafted to win votes, but fortunately there is also a strong, though tacit, recognition that in practice remuneration is often decided by the principle of supply and demand, with broadly beneficial results, and a corresponding reluctance to force into existence any specific mechanism for making remuneration 'fair'. The aim of the section is to reinforce that reluctance with rational arguments in the interest of prosperity from technology.

Technological progress is unavoidably a prime cause of unemployment and technologists have no panacea for avoiding it, but there is a special technological reason for central governments to shoulder the burden of vocational training and retraining. There is also a corollary of the cost method which would enable the level of unemployment to be reduced in some circumstances without any additional drain on the public purse. That proposal will probably be opposed if it is not ignored, but there is no obvious emotional reason for such opposition, so it might come to be considered if unemployment remains at its present obstinately high level indefinitely – as it probably will.

However there is a real sense in which the problems of unemployment are complementary to those of environmental pollution. That is to say measures to protect the environment may reduce unemployment if they are correctly devised and with that second purpose in mind. We shall consider the possibilities in the next chapter.

REFERENCES AND FOOTNOTES

1) Otto von Bismarck in conversation with Meyer von Waldeck in 1867. Also attributed to R. Butler, Conservative politician (1902-82).
2) 'Critique of the Gotha programme', 1875. The Marx-Engels-Lenin Institute in Moscow, which published *Karl Marx Selected Works* (in English) in 1942, ranked that critique with the Communist Manifesto as 'one of Marx's most important programmatic works ... [which gave] in the space of a few pages, in very concise formulation, the theoretical basis of the programme of the party of the proletariat ... [including] an analysis of the development of future communism.'
3) *State and Revolution.*

4) 250,000 prisoners are said to have been put to work on the project, of whom only 60,000 were known to be still alive when the canal was opened in 1933.
5) In his book, *Perestroika*, he wrote: 'With full cost accounting, introduced in 1987, an enterprise finances itself, its payments to the state budget being reduced accordingly', but that reform came much too late to save communism from collapse in the USSR, and in any case cost accounting with self-financing does not amount to the cost method.
6) The Independent Schools Careers Organisation and some local education authorities conduct such tests.
7) A commercial public service is one which charges its customers for what they receive, as distinct from a non-commercial public service, such as the police and fire services (in Britain) and the armed forces, which are paid for by taxation.
8) Official figures, quoted by the Institution of Electrical Engineers in their submission to the Department of Trade and Industry's Nuclear Review – 1994, indicated that half the UK reserves had already been extracted in the 15 years up to about 1989. The reserves may be much larger in reality but there is no evidence that they are comparable with the reserves of coal.
9) Report of the Independent Review of the Radio Spectrum (30-960 MHz), chairman Dr J.H.H. Merriman, July 1983, *Cmnd 9000*.
10) The allocations were 45%, 39%, 6% and 1% respectively of the spectrum between 30 and 960 MHz.
11) D. Rudd, Lecture on 'A renting system for radio spectrum?', 1985, *IEE Proceedings*, January 1986.
12) The constraints were necessary to mitigate the effects of (a) spurious emissions at harmonic frequencies, typically 3 and 5 times the base frequency, (b) mutual coupling between stations in one neighbourhood and (c) gradients and irregularities in the refractive index of the atmosphere; also to maintain the necessary spacing between the legs of two-way communications (8 MHz between the 12.5 kHz legs of VHF voice channels).
13) Some administrative problems arising in the international air traffic control agreements were tackled in the presentations.
14) 'Spectrum Management: into the 21st Century', Department of Trade and Industry June 1996. Reviewed by the author in 'Hesitant pricing of the radio spectrum' in the October 1996 issue of *Electronics World*.
15) See, for example, *Health Policy in Britain, the politics and organi-*

sation of the National Health Service by Christopher Ham, Macmillan, 3rd edition 1992.

16) *Health Policy in Britain* referred to 'the inevitability of rationing ... politicians in the past have fought shy of explicit rationing and there is no reason to suppose they will be more willing to engage in open debate about priorities in the future.'
17) In 1984 in a government-sponsored experiment. However the local population rejected it, preferring to tolerate the congestion which had led to the experiment, except in the tunnel between the island and the mainland.
18) Notably in 'A New Deal for Transport: Better for Everyone, the Government's White Paper on the Future of Transport', Department of Environment, Transport and the Regions, July 1998, Cm 3950.
19) *Tom Brown's Schooldays* by Thomas Hughes, 1857.
20) 'Fleet Street: now the truth can be told' by Bernard Levin, February 3 1986.
21) *Maverick, the life of a union rebel*, Wiedenfeld and Nicolson, 1992.
22) Report series Number 5 – 'Educational provision, educational attainment and the needs of industry: a review of research for Germany, France, Japan, the USA and Britain' by Andy Green and Hilary Steadman, *Institute of Education, University of London and National Institute of Economics and Social Research,* 1993.
23) Professor Alan Smithers at Manchester University came to much the same conclusion, after diligent research from a different starting point in his report entitled 'All *our* futures – a Revolution in British Education', December 1993.

APPENDIX – THE RADIO SPECTRUM

Radio waves are measured by their frequency – the number of waves per second – and their amplitude – a measure of their strength. Under an international convention, waves per second are called Hertz (Hz), in honour of the German physicist, Heinrich Hertz (1857–94), and, by a natural extension, kilohertz (kHz), Megahertz (MHz) and Gigahertz (GHz) refer to thousands, millions and billions of waves per second. The lengths of the waves, measured in metres from crest to crest, used to be quoted instead of their frequencies, but either measure can be converted into the other by simple arithmetic (wavelength in metres = 300 ÷ frequency in MHz or frequency in MHz = 300 ÷ wavelength in metres), so only one is necessary and nowadays the frequencies are preferred.

The term 'radio spectrum' means the whole range of frequencies which can be used for radio transmissions. However a simple wave at a single frequency can convey no useful information except that it is being transmitted. To transmit information requires a band of frequencies extending above and below that frequency and the necessary width of the band depends on the rate at which information is to be transmitted. In the range of frequencies under consideration, a two-way telephone conversation needs a band (called a 'leg') about 12½ kilohertz wide in each direction, a FM radio station needs about 150 kilohertz and a colour TV transmission some eight Megahertz (that is as much as 320 two-way telephone conversations). If any of those bands overlapped, the transmissions would interfere with one another and make reception impossible wherever a receiver was within the range of more than one such transmitter. In other words, although the radio spectrum is an abstract concept, it is nevertheless a resource without which no radio transmitter can function.

That is why the broadcasting stations are spaced out along the tuning scale of a radio set. The spacing enables the stations to function without interfering with one another, and the TV stations and all the other transmitters are separated in the same way, for the same reason. However the waves behave differently at different frequencies, which affects their usefulness for different purposes, particularly in regard to the degree to which they curve over hills into valleys and are reflected back to the ground from the ionosphere, which surrounds the earth and so transmits the waves round its curved surface by repeated reflections.

There are various ingenious devices which enable some transmitters to share a band of frequencies without much interference, but the number which can operate in any one band or range of bands is limited.

9

PROSPERITY AND THE ENVIRONMENT

The Second World War was a watershed in the history of technology, particularly in Western Europe, where it initiated major improvements in our standard of living. Before then most people worked with their hands and muscles, on farms and construction sites, in factories, quarries and mines and loading and unloading the vehicles and vessels of freight transport by land and sea. Some Western Europeans were still undernourished and cold in winter from lack of clothing and housing. For them prosperity no doubt meant having enough to eat, warm clothes and a roof over their head, but they were already a minority. The greater prosperity for which the majority aspired no doubt included better food, more clothes and a larger, more comfortable house or flat, but it also included less labour and the leisure to enjoy its fruits.

Of course the war itself caused enormous misery but for most of the survivors that was temporary. In the ensuing 50 years, those aspirations of 'the common man' (which was deemed to include both sexes) were largely realised – or at least the physical means of achieving what most people would have previously regarded as marvellous prosperity were created. Food and clothing became plentiful, houses became warmer in winter and labour-saving, the amount of muscular effort required of most workers was greatly reduced and often eliminated and millions of ordinary people acquired television sets and motor cars and started travelling extensively by land and air. The remaining barriers to that prosperity for the minority who have not achieved it in Western Europe are now social, educational and political – no longer physical. But many of those who have achieved that prosperity have raised their sights and now aspire to standards of living which may prove to be physically impossible to realise, except for a privileged few. It is not even certain that the majority will be able to maintain their existing standards throughout the next century.

That uncertainty arises from the damaging effects of human activities on the environment. Up to and for a long while after that war, such effects were barely mentioned, even by biologists. For most people 'Nature' was depicted as powerful and often cruel; 'Man' was 'pitted against its forces'; the most important difference between *homo sapiens* and the other animal species was that they adapted themselves to their environment whereas we could often adapt our environment to our needs and desires. Technologists dreamed of ever larger projects for 'taming Nature' by building roads and bridges across 'virgin territory', inaugurating global air transport, constructing immense hydroelectric and tidal power stations and even giant schemes for bringing rain to the Sahara desert by flooding the Quattara Depression from the Mediterranean Sea and damming the Bering Strait between Alaska and Siberia for hydro power and to improve the climate in that region. Fortunately some of the most grandiose of those schemes did not materialise.

A comprehensive description of the damaging effects of our activities on our environment would fill at least a book and there is no point in regurgitating the contents of the many books which have been written on the subject. The causes of the damage can be roughly classified under three headings: local, as typified by the exhaust fumes from motor cars and lorries (trucks) in towns, regional as typified by the chimney emissions from electricity generating stations and factories burning coal and oil, which cause acid rain in places remote from those sources; and global, as typified by the greenhouse gases emitted from those sources (mainly carbon dioxide) and from cattle (mainly methane). The classification is useful although some sources of pollution cause damage in more than one category; for example cars and lorries also contribute to acid rain and global warming.

There is no agreed, systematic method of bringing technology to bear on reducing the damage in any category; indeed technology is often blamed for causing it. In Chapter One, a broad principle was put forward as the foundation of such a methodology, but it will be easier to perceive its relevance if we start off with some well-known causes of serious environmental pollution and see how far we can get by common sense, as we did in Chapter Two. In that chapter the aim was to understand how technologists think, rather than to solve any of their problems (which would have required at least a book for each problem); and by the same token the aim in this chapter is to understand how societies must think in order to tackle the problems of environmental pollution. As in Chapter Two, some specific problems are included to illustrate the kind of thinking which is necessary, but optimistic readers will be disappointed if they expect a comprehensive review or packaged solution to any of them.

A preparatory exercise

Before beginning with our first problem, it will be illuminating to go back to an old problem and its solution. It was not an environmental problem but inspiration can be drawn from it in tackling the first of our selected modern pollution problems.

In the nineteenth century, sailing ships were often grossly overloaded. It was not obvious when the ships were in harbour or in calm seas, but when they reached the open seas too many were swamped in storms and sank, with the loss of the lives of many of – sometimes all – the crew and passengers, and of course their cargoes. The physical remedy was obvious: do not overload the ships in the first place; but risk analysis was unknown and human lives (especially the lives of poor sailors) were of little account. Overloading meant larger cargoes and so lower costs if the ships did not sink, and the hulls and cargoes of the ones which sank could often be covered by insurance. So when competition was fierce, the less scrupulous ship owners risked the lives of their crews to undercut the rates quoted by their more safety-minded competitors and drive them out of business.

The way out of the impasse was to pass the Merchant Shipping Act of 1876, which required every ship to have a horizontal line painted round its hull – the Plimsoll Line after Samuel Plimsoll MP, who led the campaign for the Act. The Act made it illegal to put any more cargo or passengers in any ship than would bring its Plimsoll Line down below the water level in harbour. Violations could be spotted at a glance and the owner or captain brought to book.

That act reduced the cargoes which ships could carry and hence increased the resource cost (see Chapter Three, page 86) per tonne-km of transporting freight in those ships which reached their destinations safely. One gain was that fewer hulls and cargoes were lost, but the main gain was of course that the lives of many sailors were saved. Just how the Act came to be passed is a matter for the history books. The only relevant facts in the present context are that the law was simple and not too difficult to enforce and the Act was eventually passed when public opinion became strong enough to persuade Parliament to pass it. It did not depend on an assessment of the monetary value of a sailor's life. So what is the relevance of the Plimsoll Line to modern environmental pollution?

OIL SPILLAGES BY SUPERTANKERS

Until the Second World War, oil tankers were small by modern standards. Few carried more than 16,500 tonnes of oil. If one was wrecked, the environmental consequences were not serious. Since then, the discovery of enormous quantities of cheap oil in the Middle East and elsewhere and the consequent expansion of the world consumption and sea transport of oil have led to the construction of larger and larger supertankers, able to carry 100,000 then 200,000, 400,000 and up to 500,000 tonnes of oil each.[1]

The cost of constructing and operating a supertanker to carry, say, 200,000 tonnes of oil on long voyages is of course much less than the total cost of twelve prewar 16,500-tonne tankers, which would be required to replace the supertanker, but the environmental consequences of a wrecked supertanker can be devastating. Yet they are made to less robust standards and are more difficult to steer and stop, more vulnerable when damaged by storms or mishaps and often less carefully navigated than the prewar tankers. Those are the well-known physical causes of the wreckings which have resulted in that devastation.

Until that war, the plates of a ship's hull were made of mild steel and riveted together, which made the hull slightly flexible and so less stressed in a rough sea than a fully rigid hull. The technology of the alternative method of joining steel plates – welding – was in its infancy, but it was developed during the war and enabled the Allies to build large numbers of ships very quickly to replace those sunk by enemy submarines in the North Atlantic Ocean. Modern supertankers have welded hulls with high-tensile steel plates, which are about 15% lighter than mild steel but liable to 'fatigue' cracking (appropriately named) under the persistently repeated stresses on long voyages. Moreover at sea the sheer size of these ships, and particularly their immense length, make them notoriously difficult to stop or change course except very gradually. They are in fact so unwieldy that their crews do not always bother to keep a proper lookout, relying on other, smaller ships to take the necessary action to avoid collisions.

Most supertankers have only one propeller, so that if their rudders are put out of action in a storm or by some mishap, they cannot resort to operating twin propellers in opposition in order to steer. And their anchors are too weak to bring them to rest in a shallow sea near a coast. They just drift helplessly until a salvage tug can reach them or they are wrecked. If a tug reaches a drifting supertanker, the two captains must negotiate the price at which the rescue will be made before any action is taken. Both are under pressure from their owners to make the best deal, and they may make very different judgements of the difficulties of the rescue, the likely arrival of competing salvage tugs and the dangers of further delay. If the judgement of the supertanker captain is mistaken, the supertanker may be wrecked, although it could have been rescued if the operation had commenced as soon as the first tug had reached the stricken ship.

Out of the many environmental disasters caused by supertankers in the last 30 years, five happened to hit the headlines and television screens of the Western news media. In 1967, the captain of the *Torrey Canyon*, who had tuberculosis, tried to take a short cut to Milford Haven in South Wales and could not steer clear of the Seven Stones Reef off the southwest corner of Cornwall. The ship was wrecked and 120,000 tonnes of crude oil leaked out of the hull, created havoc on the holiday beaches and killed thousands of sea birds. The second disaster was in 1978 when the *Amoco Cadiz* suffered a minor failure of its steering gear in the English Channel. Although there was a salvage tug in the locality, the captain of the *Amoco Cadiz* could not – or at least did not – agree the salvage terms offered by the captain of the tug until the supertanker had drifted perilously close to the French coast. When the terms were agreed and the salvage began, the weather worsened and the tug's tow line parted. The *Amoco Cadiz* dropped anchor but the flukes on its anchors were torn off and the ship drifted onto the Brittany shore, where it broke in two, spilling 220,000 tonnes of oil onto the beaches.

The third disaster was in 1988, when the *Exxon Valdez* ran aground off Alaska. It leaked only 36,000 tonnes of oil but in an area of special environmental sensitivity. Moreover the drama was shown live on TV screens right across the USA. Meanwhile there had been 27 larger spills in other parts of the world. The biggest was 270,000 tonnes off the coast of Trinidad in the Caribbean in 1979, but it was hardly mentioned in the American or European newspapers.

In January 1993, the tanker *Braer*, carrying 84,500 tons of light crude oil, was wrecked on the Shetland coast. In a severe storm the engines failed when the ship was mid-way between Orkney and Shetland, bound for Canada. When the ship had drifted for three and a half hours, the last of the crew were taken off by helicopter before a rescue tug could reach the scene from fifty kilometres away. Some crew were put back on board later, but they could not restart the engines nor secure a tow line from the tug. They were taken off again after the tanker had run aground in an environmentally sensitive area, six hours after the engine failure.

The tanker immediately began to spill oil, but the high winds persisted and dispersed it, so that the environmental toll was comparatively light – some 1,500 sea birds and a few seals and otters were killed. There are no polluted beaches but the island's salmon farming and tourist industries suffered financially to the tune of an estimated £50 million.

The most recent disaster to reach our TV screens was in January 1993 at Milford Haven in south-west Wales. The oil tanker *Sea Empress*, carrying 150,000 tonnes of light crude oil, made an error in its approach and went aground on some rocks. The port's four harbour tugs could not get it off in the strong tides and winds and at least half the cargo leaked out in six days until a flotilla of twelve more powerful tugs managed to refloat it. A twelve-mile long oil slick lapped a bird sanctuary and fouled 500 sea birds. Milford Haven had been regarded as a safe port but it was reported that its radar cover had been out of action for six months.

Why did these disasters happen? It is important to understand that they were not due to any insuperable technological problems. It is perfectly feasible, with existing, well-developed technology, to build supertankers which are as robust as their 16,500-tonne prewar predecessors and which can be stopped and steered as easily. The crews can keep as good a lookout now as 60 years ago and the facilities for accurate, safe navigation, taking account of currents, tides and the forecast weather (if they are in action), are better now than then.

No, the reason is entirely economic and closely parallel to the reason why sailing ships were overloaded in the nineteenth century. Any of the necessary, extra precautions to guard against disastrous spillages would increase the cost of transporting oil. The real cost (that is after allowing for inflation) would still be much less, per tonne-kilometre, than consumers all over the world were willing to pay 60 years ago, when they were mostly much poorer than they are now. But international oil transportation is a competitive – often fiercely competitive – business, so any owner or operator who took such precautions would risk being driven out of business by those of his competitors who did not.

It was perhaps plausible to argue until ten or fifteen years ago that supertanker owners and operators will be induced to adopt the necessary measures to prevent disastrous spillages if they are forced to pay full compensation for the damage they cause. But there are two counter-arguments. One is that it is very difficult, if not impossible, to assess – or even define – 'full compensation'. How thoroughly must the beaches be cleaned? How much did the holiday makers suffer? How many thousands of sea birds were killed and what were they worth in money? How many other species suffered and how disastrously – in the short term and in the long term? What is the money value of the anguish caused to present and future generations by the devastation of a large environmentally sensitive area. The report on the *Braer* disaster came up with a financial estimate, but it only covered the commercial interests and it will probably be disputed. In general such valuations are as imponderable as the valuation of sailors' lives in the nineteenth century (which is not to say that the value of sea birds can be compared with the value of sailors' lives but that both are imponderable).

The other counter-argument is that the first imperative for an operator in a competitive industry is to survive in the short term – to solve this year's (sometimes this month's) problems before worrying about next year's problems. There is no comfort in knowing that you will not have to pay compensation over the next twenty years because your ships are soundly built and scrupulously operated to avoid disastrous spillages if your careless, unscrupulous competitors undercut your rates and drive you into liquidation before they are held responsible for a disaster. One of the disadvantages of unrestricted competition is that only the short-term survivors survive in the long term. It is a trite truth but it puts a premium on taking the short-term view. For such reasons, although an enforceable obligation to pay full compensation, in so far as it can be assessed, is necessary, it has not been and is not, by itself, sufficient to induce the necessary substantial changes in the design, construction and operation of supertankers, So disastrous spillages must be expected to happen again and again until something else is done to prevent them.

The remedy, to which the example of the Plimsoll Line provides a pointer, is to have and enforce a set of simple regulations to ensure that only those supertankers which are, in a defined sense, 'environmentally safe' in their design, construction and operation can carry oil in such quantities. 'Simple' in this context does not mean as simple as the Plimsoll Line; technology is inherently more complicated now than it was then. It means simple enough to be enforced without too great administrative expense and simple enough to be understood in outline by the general public. That last condition is essential because public opinion has to be roused to persuade parliament to pass the necessary legislation. It will be difficult to rouse it if the opponents can plausibly claim that the regulations have been hatched up by bureaucrats who are more interested in advancing their own careers than preserving beaches and wild life.

Oil transportation by sea is of course an international industry, but that does not necessarily mean that the regulations must be agreed internationally before they can be enacted and enforced. In the foregoing three classes of the causes of environmental damage, oil spillages are not global incidents; they are not even regional incidents; they are local incidents with local consequences. The *Torrey Canyon* disaster did not affect the people of the USA, not even the people of France to any great extent and the black tide which inundated the beaches of Brittany in 1978 did little, if any, damage to any English beach and none at all in Italy or Scotland. The consequences of the *Braer* disaster were confined to Shetland. Also, although internationally agreed standards and regulations are generally preferable to differing local regulations, the benefits of standardisation are least when goods are produced in small numbers. In this case there are only a few hundred supertankers in the whole world and they cost of the order of £100 million each, so the economies of standardisation are minimal. Bearing in mind that international standards are notoriously difficult to negotiate, it is well worth thinking about the pros and cons of national standards.

In that connection it is noteworthy that the coast of the USA suffered little from major oil spills until 1988, but then the *Exxon Valdez* disaster was followed only two years later by the Oil Pollution Act, which required new supertankers operating in US territorial waters to have double hulls, to create a gap between the outer hull, which would probably be holed if the ship went aground, and the tanks containing the oil cargo. That act confirmed (if confirmation were needed) that a country can regulate the design of foreign ships operating in its territorial waters, without regard to the regulations – or absence of regulations – enacted by other countries.

That particular act had the disadvantage that it specified part of the detailed design of the ship rather than its performance, thus constraining the designers' ingenuity. There are other – perhaps more effective and/or cheaper – ways of limiting the amount of oil which can leak from a supertanker when it runs aground and is holed; moreover double hulls are difficult to inspect properly for rusting and fatigue fractures. It was considered that a double hull would not have given any additional protection in the circumstances of the *Braer* disaster in Shetland, when a helpless vessel was driven by storms on to rocks. In any case the old maxim that prevention is better than a cure comes to mind.

As in Chapter Seven, the principle is more important than the details – in this case the details of the necessary regulations to prevent disastrous spillages on our beaches, or at least to make them negligibly rare. Obvious performance regulations would include the ability to stop within a limited distance, and to steer round a limited turning circle when the ship's rudder is out of action. Those are easily demonstrable capabilities. An obligation to keep a continuous lookout could be monitored by flying a boldly marked aeroplane with a quiet engine across the bows of the ship at any time without prior warning and requiring the lookout on the bridge to radio the sighting on a specified frequency within a few minutes, under pain of a large fine if the sighting were missed. It could be insisted that the captain and crew should have regular medical examinations and the supertanker's course be registered in advance. A set of contingency rescue contracts, deposited with the registration authority and covering the whole course, might be a back-up requirement. Tankers could be barred from entering a port whose tugs are too few or too small or whoses radar installation is out of action. And so on, but not *ad infinitum*.

Performance and operational requirements of that nature have the important virtue that they give the fullest scope to the designers and operators to comply at the lowest cost. Operating twin propellers in opposition under the stern is not the only way of steering a ship when its rudder is out of action; a small reversible transverse propeller in the bow, driven by an electric motor, could swing the bow in either direction more easily and probably more cheaply. And there are several ways of bringing a ship to rest[2].

There can be no insuperable enforcement difficulty if it is kept in mind that there are only a few hundred supertankers in the world. Keeping track of them and, if necessary, extracting fines from their owners or operators, when they next call at a national port, would be a straightforward business in view of the very high value of the ships and their cargoes.

It may be objected that such national regulations would not, by themselves, have prevented the *Braer* disaster because the *Braer* was not sailing from nor bound for a British port and it got into difficulties while in international waters. That sort of objection can be made to any new regulation (including the Plimsoll line, which could not save the lives of sailors in ships which were not subject to that kind of regulation), but it neglects the beneficial consequences of setting an example. Hard on the heels of the American Oil Pollution Act, the International Maritime Organisation, the United Nations body which governs world shipping, introduced new rules under which new tankers built after July 1993 have double hulls and all tankers, new and old, had to have them by July 1995. It is arguable that those rules are less cost-effective in reducing oil pollution than a set of simple rules about stopping, steering and keeping a proper lookout from the bridge, but that is not a compelling reason for resisting the construction of double-hulled vessels now that they are obligatory.

We seem to have made substantial progress with our first pollution problem by combining a knowledge of technology with common sense, which so notably failed in Chapter Two. But in the process we have brought the notion, suggested in Chapter One, of pleasant surroundings, or the avoidance of unpleasant surroundings, into what we are trying to achieve. Oily beaches are frightfully nasty, however little some holiday makers may care about the sight and sounds of the sea birds; moreover many of us care very much and by now we are also afraid that the future consequences of wiping out hundreds of species of flora and fauna will be dire, even if we still do not understand why. So how can the case be presented to the general public?

Before the Second World War and for many years after it, our beaches were unsullied by oil but most people could not afford to visit them for their annual holidays, or they could go to only a few beaches near where they worked. Now that many more people have motor cars, millions of us drive to even the most remote beaches every year, using petrol from the cheap oil which is transported in supertankers, but sometimes our holidays are ruined by oil spillages from supertankers, which are less well designed and operated (from the point of view of avoiding oil spillages) than prewar ships. At the cost of a small increase in the price of that petrol, those spillages could nearly all be avoided. We holidaymakers could then enjoy ourselves to the full and return home without the nagging fear that our quality of life is being slowly undermined by the disastrous consequences for other species of allowing largely unregulated operation of supertankers.

But no motorist on holiday can choose individually between cheap petrol but a possibly oily beach on one hand and slightly dearer petrol but an unsullied beach on the other. It is not like buying groceries, where one shopper can choose the cheapest tea and another can pay more for a better quality. The only way of preventing disastrous oil spillages is for the voters to pester their members of parliament in such numbers that the necessary regulations are brought in. Ship designers all over the world will then bend their brains with enthusiasm to meeting the regulations at the lowest cost, that is the smallest increase in the price of petrol. The technology and the necessary features in society to direct it towards prosperity in that enhanced sense are already established in the democratic countries. If the voters cannot be roused to play their part, it has to be regretfully concluded that by and large they prefer very cheap petrol to unsullied beaches.

That is a slightly simplified presentation. Britain produces about as much oil as it consumes in total, but we export some grades, of which we produce a surplus, and import others, of which we do not produce enough for our consumption. So the regulations would affect petrol prices in Britain in a complicated way which would make it more difficult to calculate the extra cost to motorists than might appear at first sight. Opponents of the legislation – or some of them – could be expected to exaggerate the cost of compliance, just as did the opponents of the Plimsoll Line, but even if their initial estimates had to be accepted, the cost would fall when the competing designers and operators got to grips with the problems. No legislation will be enacted in the field of environmental protection if it depends on precise calculations of the ultimate costs.

AIR POLLUTION FROM ROAD TRANSPORT

Our next example is chosen partly to dispel any possible illusion that common sense is always a reliable guide to the solution of pollution problems. Electric road vehicles with batteries (EVs for short) were used in Chapter Three to illustrate some facets of the cost method in designing applications of technology. Let us now return to them as possible means of reducing local air pollution.

In 1985, small electric passenger cars and vans able to carry up to one tonne or so of freight could be purchased in small numbers, and it was estimated that in Western Europe up to 6 million small cars and one million light vans, on short journeys in towns, could feasibly have been replaced by EVs if costs were temporarily ignored. The following four disadvantages and five advantages of those EVs, compared with their counterparts with internal combustion engines (ICVs), were given in that chapter (pages 53–4):

1) a reliable battery was about 150 times as heavy as a tank of petrol or diesel fuel which would give an ICV the same range as the EV before the tank had to be refilled or the battery recharged;
2) recharging the batteries took several hours rather than the few minutes needed to refill a fuel tank;
3) the batteries were expensive and wore out before the vehicles;
4) the vehicles themselves, without batteries, cost three to five times as much as ICVs with equivalent capacity and performance.

On the other hand:

1) the EVs were much quieter;
2) they did not pollute the air in towns;
3) lighter, potentially cheaper batteries with estimated useful lives of up to four or five years had been developed and more improvements were expected;
4) both the vehicles and the batteries would be much cheaper if they were produced in large numbers;
5) the EVs themselves did not require for propulsion any product of petroleum, of which Western Europe was a net importer, because electricity can be generated from several different sources of primary energy, as mentioned in Chapter Two.

Thirteen years later the position is much the same – indeed it has not changed really radically in the last 50 years (some would say since ICVs were invented at the turn of the twentieth century). The batteries are still too heavy, expensive, slow to recharge and short-lived and the resource costs (excluding subsidies) of the EVs without batteries are still much higher than the cost of equivalent ICVs. On the other side, the EVs are still quieter (though not so much quieter) and less polluting (though not so much less polluting) than ICVs, they would still be cheaper than they are if they were produced in larger quantities and they still do not depend on any product of petroleum for propulsion, of which Western Europe is still a net importer (though its deficit has diminished). Moreover, although the improved batteries have still not materialised in commercial quantities, the manufacturers and other interested parties still aver that they will in another few years. Sceptics have said for many years that the commercial horizon above which the EV enthusiasts believe their vehicles will rise is always five years hence. One wonders how such optimism – and the funds for more research and development which it generates – can persist for so long.

The simple explanation is that many people, even technologists, just assume, without enough thought, that the quietness of EVs and the absence of any direct gaseous emissions from them must make them ideal in some sense for urban transport, in spite of those four disadvantages. It seems obvious to their common sense. But that is because they look at the problem from the wrong end. Let us examine the two 'environmental' advantages.

Noise

Internal combustion engines are inherently noisy because combustion is a chaotic process, especially in a confined space such as a cylinder of an internal combustion engine. In the absence of counter-measures, the noise which comes out of the exhaust pipe is shattering. In an EV on the other hand, the combustion is banished to a remote electricity generating station, where it is less confined and only a few people have to put up with it; moreover the ones who have to endure the worst noise (in the boiler house) are paid to work there and they can wear ear muffs if they can be bothered to put them on. And of course if the generating stations are hydro-electric or nuclear, there is no combustion noise, although the other processes make some noise, as do the EVs themselves.

A survey of the effects of road traffic on the environment of homes in 1983 found that noise from road traffic affected people at home more seriously than noise from any other source[3]. Safety was the only factor which the respondents in the survey thought was more important than noise.

The noise produced by the engine and its accessories in an ICV is called 'power train noise' and reducing it to the point where the ICV would emit no more noise than an equivalent EV would be very expensive. But the human ear is so adapted that two equally noisy vehicles passing simultaneously sound much less than twice as loud as one of them by itself and a hundred such vehicles sound very much less than ten times as loud as ten. It follows that removing a small number of noisy vehicles from a heavy stream of traffic and replacing them with quiet vehicles would make little or no difference to the noise level[4].

Moreover the noise level in an urban street is determined mainly by the noisier vehicles, even if there are not very many of them among many quieter vehicles. The worst traffic noise in most towns came – and still comes – overwhelmingly from heavy vehicles, particularly long-distance lorries (trucks), which are much noisier than cars and vans on urban roads but which cannot feasibly be replaced by EVs, mainly because of their necessarily long journeys. The benefit of reducing the noise of the small cars and vans would be imperceptible until: (1) the noise of the lorries – or most of them – had been reduced and (2) the proportion of the quieter small vehicles in the total traffic was much larger than the numbers which could feasibly be replaced by EVs would represent.

EVs might make an effective contribution to noise reduction in some special places, notably pedestrian precincts, selected city centres and residential areas where there is little traffic and from which heavy, noisy vehicles are already barred, also in rural districts into which noisy lorries seldom have occasion to go. However those places are much quieter (without any EVs) than the unrestricted streets where many people still live and/or work but there is a lot of traffic and heavy lorries are allowed. Introducing EVs in restricted areas would leave the major problem to be dealt with by other means. The EVs would be like plasters on a cut finger when the patient also has a broken arm.

When the international team, which was studying the prospects for EVs for its members' governments as described in Chapter Three (page 52), understood that implication, it wisely forbore to emphasise the quietness of EVs as an important determinant of those prospects.

Exhaust gases

The potential physical benefits of EVs in reducing pollution from the exhaust gases of ICVs are less elusive. The main pollutants are carbon monoxide, nitrogen oxides, hydrocarbons, sulphur dioxide and harmful solid particles. Two similar ICVs emit twice as much of those pollutants as one and a hundred emit ten times as much as ten. EVs emit none but EVs are the indirect cause of nitrogen oxides, sulphur dioxide and particle emissions from the generating stations which supply the EVs if the stations burn coal or oil. Those emissions are far less damaging than the emissions from the ICVs, mainly because they come into less contact with people, but they cannot be left out of account. (ICVs and EVs also both increase the global emission of carbon dioxide, which is a 'greenhouse' gas to be considered in a later section of this chapter, but there was not enough difference between their emissions to have any appreciable effect in the 1985 comparison.)

In 1985 the study team estimated that the six million small cars and one million vans, distributed between the ten participating countries, could reduce urban air pollution by as much as 20–30% in two countries (the then Federal Republic of Germany and Britain), as examples of their potential benefit; and corresponding estimates followed for the other eight countries. The net effects on total emissions from all sources were estimated to be much less in those two countries because of the emissions from the generating stations, but the net effect would be beneficial. Moreover in countries which burned little or no coal or residual oil in their generating stations the harmful effect of electricity generation would be negligible.

The difficult question was how to weigh those undoubted advantages – which were measured in tonnes of the pollutants – against the higher annual costs of the EVs – which were measured in European Currency Units. Those costs could be brought down by manufacturing the EVs in the quantities envisaged, but they would still be higher than the costs of equivalent ICVs on the same missions. It was no use just stating the advantages and disadvantages of the EVs in their different units because that sort of thing had been done before. It was essential to present a balance sheet with the items on both sides in the same units – which had to be European Currency Units.

Some previous studies had attempted to calculate the losses of income and the costs of treating the affected people, animals and plants and repairing the damage to property which actually occur in towns and are caused by ICVs, but their results were not highly regarded. To repeat them and bring them up to date would have meant collecting a mass of technical, traffic-related, town-planning and climatic data and the results of physical, chemical, toxicological and medical research, and then trying to devise a formula to express the financial cost of the damage to the health, welfare and property of the town dwellers in terms of the quantities of those pollutants. To make the task even more difficult, the formula would have had to express not just the total cost in terms of the total emissions of all the ICVs but the reductions in that cost which moderate, variable reductions in those emissions would bring about in the numerous different locations.

The study team concluded, with some regret by most members, that it was 'not possible to meet those requirements in the framework' of the study – and so turned to another method. The author believed – and still believes – that such calculations are not merely impossible in a limited framework but totally impossible in any framework – for the reason that, although a hundred similar ICVs on the same mission emit ten times as much of those pollutants as ten, it cannot be said that they cause ten times as much lung cancer or severe asthma or that they kill trees and corrode buildings ten times as fast. They may do twice as much damage or a hundred times as much – if indeed diseased lungs, dead trees and corroded buildings can be added together to get an answer in ECUs, which is itself a doubtful proposition. And there was an additional reason (which we shall come to) for turning to the other method.

That other method arose from the circumstances of the time. Research and development over many years had yielded two means of reducing the polluting emissions from petrol engines, each with some advantages and disadvantages. The first was to adapt the engines to take in more excess air, so ensuring that the combustion was more complete, albeit with some loss of performance. Those engines, known as 'lean-burn' engines, emitted very little carbon monoxide or hydrocarbons (the products of incomplete combustion), but unfortunately the higher combustion temperatures increased their emissions of nitrogen oxides. Mainly for that reason, lean-burn engines were not brought into commercial production.

The other means, which has since become mandatory in new vehicles, was to put a so-called 'three-way catalytic converter' in the exhaust pipe. Two catalysts complete the combustion of the carbon monoxide and hydrocarbons to produce relatively harmless carbon dioxide and water vapour and the third decomposes the nitrogen oxides. The disadvantages (apart from the cost of the converters) were – and still are – that they did not operate until the engines were warm and so were ineffective or less effective on short journeys; and they increased the petrol consumption by about five per cent. It had to be lead-free petrol, which had a higher resource cost than leaded petrol, but leaded petrol was already on the way out and so was not counted against the petrol engines in the balance sheet.

Catalytic converters were expected to reduce the emissions of those three pollutants by up to 90% on long journeys – much less on short journeys. So they were less effective per vehicle than EVs but that deficit could be corrected in the calculations because, as already mentioned, only a limited number of ICVs could feasibly be replaced by EVs, taking account of the daily utilisations and missions of the vehicles in relation to the short ranges of the EVs and the time taken to recharge their batteries.

The differences between the advantages of the zero emissions of EVs and the reduced emissions of petrol vehicles with catalytic converters were allowed for by comparing the annual costs of fleets of EVs with the costs of larger fleets of those petrol vehicles, such that the fleets made equal contributions to reducing air pollution. On average about 1,500 ICVs would have had to be fitted with converters to reduce the pollution by as much as 1,000 EVs, compared with ICVs without converters in each case. That probably sounds complicated and it was, but it depended solely on the costs of real equipment and processes, namely the converters and the flue-gas cleaning process in the generating stations, not on impossible estimates and subjective judgements of the benefits of reducing air pollution by, say, 25% in one town, 13% in another and so on right across Western Europe.

The corresponding estimates for the advantages of replacing light diesel vans by EVs were more speculative. Diesel engines, when correctly adjusted, emit little or no carbon monoxide or hydrocarbons, but they work at higher combustion temperatures than petrol engines and so emit more nitrogen oxides, plus some sulphur oxides and harmful particles which petrol engines do not emit. There was no means of substantially reducing those emissions, as it were 'waiting in the wings', but there was some evidence that they could be reduced if the general public came to think that it was necessary. So it was assumed that it would be possible to reduce the emissions of diesel engines to the same extent as petrol engines and at the same cost per vehicle. That was a rather dubious assumption at the time but subsequent developments have underpinned it to some extent.

Doubts, criticisms and confusion

Some EV enthusiasts argued then – and would still argue today – that the study should have credited the EVs with all the benefits – to health, plants and property – of the lower levels of pollution which they would bring about. They brushed aside the difficulties, saying that it was misleading to conceal the fact – as they saw it – that those benefits would be far larger than the penalties which were ranged against the ICVs in the comparisons.

The counter-argument, which won the day, was that the benefits would come – and indeed they are now coming – from the lower levels of pollution, and the EVs were just one way of achieving those lower levels. In general all our decisions in life, whether as individuals, organisations or communities, are choices between available options. Thus when one buys a refrigerator one chooses between the available means of storing food at low temperature, normally between different makes of refrigerator. In a long emergency, such as a world war, one might have to choose between just one make of refrigerator and having to manage without any cold storage for food, and in that case the value of the refrigerator (what one would be willing to pay for it if one could possibly afford it) would be much higher than the non-emergency price, but that is not the normal situation.

By the same token, the value of EVs would have been much higher if they had been the only means of reducing air pollution in towns, but they were not the only means. They could not eliminate pollution altogether, even in towns, because it was not technically feasible to substitute them for all the ICVs in towns. They could make only a limited reduction and there were other ways of doing that; so the choice was between EVs and those other means. That was the additional reason (see page 354) for turning to the other method of evaluating the benefits of EVs.

That controversy illustrated how EV enthusiasts habitually look at the problem from the wrong end and so mislead the general public, who cannot spare the time and effort to examine in detail all the problems which societies must face. The real problem in this case was not and is not: how can we find a niche for EVs in our transport system? It was and is: how can we best reduce urban air pollution to acceptable levels? EVs might have made a small contribution but another way of reducing air pollution by far more than EVs could achieve was being pursued in parallel and was adopted.

We should remember that 40 odd years ago, the Clean Air Act, 1956 and the equivalent legislation in other countries brought about enormous improvements. The compulsory fitting of catalytic converters in petrol vehicles is now making a further large improvement and probably the exhaust emissions of new diesel vehicles will be cleaned up to much the same extent in the next few years. The effects of those measures will be felt only gradually as existing vehicles are replaced.

There will be those who protest that that is much too slow, but EVs would not be an effective means of speeding up the process. The obvious way of doing that would be to phase out the old ICVs more quickly. Future EVs in Europe will have to be compared with ICVs which have been cleaned up to the extent which is commercially feasible at that future date, not with the aging generation of heavily polluting ICVs which are at present (1997) still allowed to operate.

In such a future situation, which can be foreseen with enough confidence to be worth thinking about, the relevant question will be how much the general public, when correctly informed of the options, will want to make any further reduction of air pollution. What will people be willing to do without in order to achieve that further improvement?

The Californian strategy

Until 1995 the California Air resources Board had a strategy to deal with that situation. Over a million new cars are sold in California every year and the intention was to require a two per cent quota of every major car maker's sales in 1998 and thereafter to be zero-emission vehicles, with a fine proportional to the number of such vehicles by which the maker's annual sales fell short of its quota and with possible exclusion from the Californian market as a back-up penalty if the maker failed to comply. The quota was set to rise to ten per cent by 2003, and some other states were set to follow suit. Hydrogen-fuelled vehicles counted as 'zero-emission', but they were and still are even further over the commercial horizon than EVs, so the legislation would have amounted in practice to forcing the manufacturers to make and sell EVs in those numbers.

The makers estimated they would cost more than twice as much as an ordinary family car and the only batteries yet commercially available would give them a range (variously reported) of 110–160 km (70–100 miles) when the batteries were fully charged, which is not much more than the mileage left when most motorists, driving ordinary cars, would be beginning to think about pulling in for a refill. That would be a daunting prospect for the EV drivers, who would also know that, when they did pull in, they would not be able to fill up in a few minutes. If they had not reached their intended destinations for that day, or they had a change of mind which took them beyond that range, they would have to wait for their batteries to be recharged before they could complete their journeys.

A few such EVs were expected to go to the electricity generation utilities and some other concerns, which were assumed to have a financial interest in EVs and a strong enough belief in their future to pay the higher first cost and put up with the inconvenience and higher operating cost caused by their short range. But there was no reason to suppose that ordinary motorists would buy EVs unless they were subsidised by the makers to offer them at no more than the same prices as ordinary cars, and probably less to compensate for that short range and the waiting time for recharging the batteries.

The Californian EV enthusiasts put the extra first cost of the EVs at no more than 20% over the cost of ordinary family cars. For comparison, the European EV makers who co-operated with the 1985 study provided the following estimates for two electric passenger cars, compared with their nearest equivalent petrol cars:

	FIRST CAR		SECOND CAR	
	electric	petrol	electric	petrol
Carrying capacity kg or passengers	330	515	4	5
Maximum speed km/hr	100	150	55	115
Time to reach max. speed from rest secs	13	5.5	>20	7
Range of EV km	60		55	
Cost ratio (EV/ICV), including batteries, for annual production of 10,000 vehicles	3.1		1.6	

The cost ratios were estimated to fall to 2.0 and 1.5 for an annual production rate of 100,000 vehicles, but 2% of the 1 million cars sold annually in California is only 20,000, to be shared between several makers, so 10,000 is more realistic than 100,000.

It can be seen that the expected performances of the European EVs were substantially poorer than their petrol counterparts and their ranges substantially less than had been stated for the Californian EVs. Moreover the European companies did not undertake actually to make any EVs for sale at those prices and were criticised for allegedly pitching their estimates too low in order to persuade their governments to sponsor the development of EVs. All in all, although the American makers might be suspected of exaggerating the costs of EVs, their estimates appeared to be more realistic than those of the EV enthusiasts. The fact that the car companies developed catalytic converters under pressure of new control standards adopted in 1969 is not a good parallel in as much as those standards applied to all new cars, so the makers did not have to sell the converters to sell their cars.

As a working hypothesis, let us suppose that the makers would be able to sell their quotas if they offered them at the same prices as ordinary cars. Then to cover their higher estimated manufacturing costs of the EVs, the makers would have to increase the prices of their ordinary cars by 2%. If the EVs travelled the same annual mileages as the ordinary cars, they would reduce the pollution by 2%. In practice they would be bought mainly by families with two or more cars, who would continue to use their ordinary cars for their longer trips and relegate the EVs to short trips for which their range was adequate. That would reduce their contribution to pollution reduction, but those shorter trips in the EVs were likely to be in the towns, where the pollution is worst. There were too few data to calculate the reduction of pollution with any accuracy, but it would obviously be very small – as would the extra cost.

The same hypothesis applied to 10% of EVs in 2003 (still only 100,000 a year shared between several makers) would show a 5–10% increase in the prices of ordinary cars to produce a 10% reduction in pollution. But the EV enthusiasts argued that by then the new improved batteries, which they said were just over the horizon, would be commercially available and would give the vehicles much longer ranges – long enough to make them much more useful to the motorists who would be expected to buy them and correspondingly better able to contribute to pollution reduction.

But to anyone who was not an EV enthusiast and not employed by a battery or EV manufacturer but was aware of the history of battery research throughout the twentieth century, especially during the second half of it, those expectations sounded like wishful thinking. The author was the project manager of a British government sponsored research programme on sodium-sulphur batteries, which are much lighter than conventional batteries for EVs but have to work at high temperature, in the late 1970s, but 20 years later they are still very far from commercial availability. And there have been half a dozen other advanced battery research programmes in Europe, the USA and Japan which have similarly run into the sand[5].

The California Air Resources Board gave up the first part of its zero-emission strategy at the end of 1995. The Board retained the second part (requiring 10% of new cars to be

ZEVs in 2003), but it was evident that, short of a totally unexpected research breakthrough, that requirement would not be met.

That does not necessarily mean that further research into batteries for EVs should be cut off. Allocating funds for technological research always involves decisions about priorities, which are almost never tidy and often painful. Research is like that. The mistake in this case (and it cannot really be described in more diplomatic language with any hope of it being understood) was that the legislators, or their advisors, were still looking at the problem from the wrong end. They still assumed, without enough thought, that the quietness of EVs and particularly the absence of any direct gaseous emissions from them must make them ideal in some sense for urban transport, in spite of their known disadvantages.

Arising probably from that mistaken assumption, the Californian legislators, like the European EV enthusiasts in 1985, paid insufficient attention to two vital points, namely:

1) Pollution comes not from possessing an ICV but from driving it, and the quantity of pollution is proportional to the mileage driven, so it will not help to make EVs available at competitive prices without a strong inducement to reducing the annual mileages of the ICVs;
2) Although the designs of EVs have improved since the turn of the century and since 1985, so have the designs of ICVs. For a valid comparison of alternative technologies, what is available or expected to become available in one must be compared with what is available or expected to become available in the other at the same time, not overlooking that if one technology is expected to progress over the next five or ten years, so may the other. At the time of writing the latest improvement in reducing pollution from ICVs is reformulated petrol, which is reported to have dramatically cut traffic pollution in four states in the USA.

In the eyes of a technological realist the Californian strategy was doomed from the start. We need a more thoroughgoing approach, for which we must consider the nature of decisions about protecting the environment, rather as we considered the nature of technological decisions in Chapter Two.

THE NATURE OF DECISIONS ABOUT PROTECTING THE ENVIRONMENT

If we were candidates in an examination on Environmental Studies, how should we answer the question: 'Compare and contrast how individuals and small groups take large financial decisions with how societies take – or should take – decisions about protecting their environment?' Or, if you prefer, how would you mark the paper of a candidate who answered as follows:

When, as individuals or families, we are wondering whether to look for a house to buy rather than continue renting a flat or living with our parents, or to look for a car to buy rather than travel by public transport, our reasoning cannot be precise in the early stages. On one side are the expected costs (capital and operating costs and charges) of the house or car, compared with the rent or fares, and on the other our much vaguer, though equally serious, feelings of how much we want them and whether we can afford them.

Few people try to calculate the money value to themselves or their families of a new house or car numerically, and in the author's opinion those few deceive themselves; their calculations are illusory. Most of us try to ascertain the cost as best we can and then mull over whether we can afford to pay as much as or more than that. Sometimes it is obvious that we are able and willing to pay more than that, so we go ahead, or not as much, so we abandon the project; sometimes we cannot make up our minds and agonise for months. In that sense, decisions about reducing pollution are broadly similar. The costs can usually be approximately estimated but the benefits cannot be calculated in money.

In another sense, however, the decision processes are quite different. Individuals or small groups, such as families, can decide for themselves whether or not to buy a house or a car without worrying about the rest of the community because the interests of the community in individual purchases of houses and cars are built into the legal constraints on the parties involved – such as the laws of contract, the building regulations, the planning laws, the obligation to pay Council Tax, VAT and fuel tax, the motor vehicle construction and use regulations, the speed limits and so on. Decisions about reducing pollution, on the other hand, have to be taken by (or at least on behalf of) the communities whose interests are going to be affected thereby. And of course communities implement their decisions about pollution by increasing (or sometimes relaxing) those constraints and enforcing compliance with them. The problem is that it is much more difficult for a whole community to come to decisions than for an individual or family to do so. (One hears of families who cannot agree about anything but presumably they are exceptional.)

Let us return to Chapter One for a moment. Reference was made there (page 13) to a lake which would become heavily polluted unless and until the community on its shores included the condition of the lake in its concept of prosperity. The condition of the air in a town is a close parallel and the first requirement for reducing air pollution in a town is for the inhabitants to include the condition of the air in their concept of prosperity, just as most of them already include the possession of a car and the freedom to use it, also the freedom to buy goods which are conveyed to the town in heavy lorries.

Chapter One also posed the question whether the features to be derived in the later chapters, particularly the emphasis on costs, must be dismantled in order to protect the environment. It is sometimes argued – or just assumed – that costs are sordid and that societies which rely on them for their really important decisions will inevitably decline. The counter-argument, which has been stated in Chapter Two and again in Chapter Eight but is often dismissed without consideration, is that the cost method is the only method by which technologists can take rational decisions. So technologists will go on using the cost method, whatever anyone may say, except in so far as they are persuaded or physically or legally restrained from doing so.

Technologists depend on elaborate structures of costs and prices for those decisions. Societies which lack that kind of structure cannot prosper until they have built one up, and if a society which has such a structure starts to dismantle it, that society will really decline – to the point of total collapse if the dismantling is not checked. And it is no good shaking one's head and muttering that there must be a better way. Those who challenge the cost method must invent a genuine replacement for it or give up the challenge – and no-one has come anywhere near doing the former. So the way forward – the only way forward – is to modify and extend the physical and legal constraints within which the designers operate so that those constraints incorporate our revised concept of prosperity, which means refining that concept until it can determine what those constraints should be.

That refinement is inherently difficult, partly because large numbers of people have to consent to it but also because the effects of our activities on the environment are slow. When we buy a house or a car, we start to feel the benefits immediately or soon – the space and comfort of the house, the speed and comfort of the car and the freedom to do what we like in the house and go where we like in the car. On the other hand when our cars are fitted with catalytic converters in their exhaust pipes and the other measures against pollution are taken, the benefits come so slowly that we hardly notice them. After some years the air begins to smell sweeter, we suffer less from bronchitis and asthma, we do not contract cancer or we live longer before contracting it, our children suffer from those diseases much less and they have more energy and more enterprise.

It requires a conscious effort of memory, imagination and will to link those benefits with the catalytic converters which were fitted ten or twenty years earlier and with the public transport on which we travelled with such initial reluctance. How best to rouse that effort on a sufficient scale is not within the scope of this book, but if and when it can be mobilised, how should it be guided to yield the best results at the lowest cost?

THE COST METHOD IN PROTECTING THE ENVIRONMENT

The first step is to distinguish between pollution which can be wholly prevented or reduced to an acceptably low level, at some cost but without wrecking the functioning of our economy, and pollution which cannot be prevented on those terms but can be reduced.

Oil spillages from supertankers are in the first category. The risk of disastrous spillages could be reduced to an acceptably low level by the measures outlined earlier in this chapter (page 344). Their enactment depends only on the will of the legislature, which depends in a democracy on the growth and emergence of a public consensus that we can afford the cost, as reflected in the higher prices of petrol and diesel oil and the other products of oil refineries, in order to prevent the spillages, which we regard as detrimental to our prosperity. Those higher costs are susceptible of estimation, albeit only within rather wide limits of accuracy because the relevant expertise is largely in the hands of the shippers and ship builders, who are interested parties, but the difficulties are not so great as to require any radical change in our habits of mind – except the habit of disregarding the effects of our activities on our environment.

Road vehicles with petrol and diesel engines are mainly (though not wholly) in the second category. They are now such an important part of our lifestyle that we cannot contemplate dispensing with them; our economy would be wrecked if we tried to do so and we should have to rebuild it (with probable further damage to the environment) before we could make any further progress. But nor can we have that lifestyle in the present state of technology without the pollution which it causes. So some air pollution in towns is going to continue well into the future. The question is how best to reduce it, if and when the collective will to do so can be mobilised.

A small reduction might be achicvcd by exhortation, but it has been tried over many years with inadequate results The straightforward way of reducing the demands for petrol and diesel oil is to levy taxes on them. Unfortunately our instinctive response to any proposal for a new tax or a tax increase is to fear that it will reduce our prosperity, but, as any reputable economist will confirm, it is not bound to reduce our collective financial wealth. Taxes are transfer payments from the tax payers, via the Treasury, to the recipients of the Treasury's disbursements. Apart from the costs of collection, administration and disbursement, the losses by the tax payers are equal to the gains of the recipients, which means that the net direct financial effects of taxation on a community as a whole are very small. The purpose of this kind of taxation lies not in its direct effects but in its indirect effects on the tax payers. So, apart from minimising those costs of collection, administration and disbursement, the important things are to ensure that the indirect effects are beneficial, rather than harmful, and that the tax payers can be persuaded to accept them, although they will not all individually benefit as much from the disbursements as they pay in the taxes.

The classic example of a harmful tax in the twentieth century in Britain was the horsepower tax on cars between the two world wars, which was described in Chapter Three (Appendix 2). Its indirect effects were to increase the manufacturing and maintenance costs of British car engines, shorten their lives and make them less attractive in the export markets. Taxes on petrol and diesel oil are not harmful in that way, but it must be remembered that it is not road transport *per se* which pollutes the air but the exhaust fumes from the petrol and diesel vehicles. EVs would be encouraged by higher petrol and diesel taxes, with a beneficial effect in reducing pollution, but cars with catalytic converters would not because they consume more fuel than otherwise similar cars without converters, although they pollute the air much less.

Fortunately the cost of catalytic converters is much too small to wreck our economy or seriously discommode motorists who are required to have them. So they are in the first category; they can be made compulsory, and largely they have been. By the same token, better designs, which operate more effectively on short journeys, should be made compulsory as soon as they become commercially available. Then the main effect of petrol tax is and will be that people and organisations which desire or require to use ICVs on roads will curb their demands for them to the extent to which they find it financially necessary or economical to do so, thereby reducing the total amount of pollution from exhaust fumes which would otherwise occur over the whole country.

The phrase 'which would otherwise occur' is important in as much as the response to higher taxes is inclined to be very sluggish. People are generally reluctant to change their habits, especially in many cases their motoring habits, and many are tied to those habits by financial necessity – that is to say by secondary needs, as described in Chapter Eight (page 308) – if for instance they have taken a job in a distant town and there is no other way of travelling to and from it. However, provided everyone expects the tax to be permanent, its long-term effects must be to dissuade people from committing themselves to regular long journeys unless they judge the benefits to them to be greater than the commercial cost, including the higher taxes, and to persuade people who are already so committed to look for ways of becoming less dependent on long journeys. A gradual but unremitting increase in the real rate of tax is fairer and more effective than a sudden jump.

However a refinement is necessary because air pollution in towns is generally more deleterious than the same quantity of air pollution in rural areas, where the population density is lower so fewer people suffer; also the traffic density is less and the wind blows more freely to disperse the fumes. Ideally, therefore, the taxes on petrol and diesel vehicles should be at higher rates per kilometre in towns than in the country, but fuel taxes could not be made to differentiate in that way. Some form of road pricing, as described in Chapter Eight (page 311) could do so. In that chapter the purpose was to allocate a scarce resource, *viz.* urban road, to the best advantage between competing vehicles. Now the same physical equipment and collecting arrangements could be used to reduce air pollution in towns, without penalising country dwellers for a degree of pollution which their activities do not cause and from which they suffer much less than town dwellers.

It is that reduction of pollution which constitutes the environmental justification of higher fuel taxes and road pricing. It is additional to the justification which was given in Chapter Eight. The benefit would be greatest in towns and would be paid for by those who wished to continue using ICVs there and found it economical to do so. Country dwellers would pay less and receive less benefit. EVs could be charged at lower rates per kilometre than ICVs because EVs do not pollute the air, but they do contribute to congestion – or they would if there were many more of them.

Unemployment

The section on unemployment in the last chapter drew attention to the two non-military threats to existing prosperity – an aging population and damage to the environment – but remarked that at least they will also help to reduce unemployment, unless they are very badly handled. If that connection was not apparent then, it may be clearer now.

The additional employment comes from such activities as making and fitting the catalytic converters in our cars, building better-designed and operated supertankers or increasing our exports in order to buy them or lease them from suppliers overseas – in fact doing all the things which are necessary to protect our environment. In that connection it is important to keep in mind that the causes of our impoverishment are those of our activities which pollute our environment. The measures we take to protect and improve it do not impoverish us; they increase our prosperity by providing or restoring a pleasant way of life. If we can be bothered to take an interest in macroeconomic indicators and are willing to believe the analysts who study them, we shall learn that measures to protect and improve our environment also increase our gross domestic product.

Unemployment and environmental pollution are two gloomy subjects. It may help to lighten the gloom a little if we remind ourselves from time to time that, if we are careful to take account of the technologies involved, the unused resources represented by the former can be used to tackle the latter.

ACID RAIN

'Acid rain' is a useful piece of journalistic shorthand for a complicated chemical and biochemical phenomenon, which is best known for the damage it does to lakes and forests. It comes from the sulphur dioxide which is formed by the combustion of coal and oil if they contain sulphur as an impurity – which unfortunately they usually do – and from the nitrogen oxides which are formed when any fuel is burned at high temperature in air, because air is composed mainly of nitrogen and oxygen. 'Rain' is strictly a misnomer in as much as the pollution is often deposited as solid particles, which then dissolve in whatever moisture they next come into contact with, but the effect is the same as if they were dissolved in rain – an increase in the acidity of the moisture on the surfaces of the plants and of the ground water in that vicinity.

Electricity generating stations burning coal and oil and the internal combustion engines of cars and lorries (trucks) are the main sources of acid rain. The damage does not normally occur near those sources but hundreds of kilometres away, which classifies the pollution as regional. No country can solve it without the co-operation of the other countries in the same region.

Some lakes are not irrevocably harmed because they have natural reserves of alkalinity, but in many cases the increase in acidity is enough to kill the fish and otherwise upset the balance of aquatic life. Whether anyone fully understands all the chemical and biochemical reactions in forests is doubtful; they are certainly too complicated to be described here. It seems that coniferous forests are more susceptible than deciduous ones and that the worst havoc occurs at high altitudes during long sunny winters with little wind. The acidic clouds, which have come from the generating stations and vehicles, hang over the foliage for months on end and the strong ultra-violet rays in the sunshine at those high altitudes cause complicated reactions in which the acids interfere with the metabolism of the trees and destroy the foliage. About one third of the German Black Forest, which covers some 7,000 square kilometres, much of it higher than 1000 metres above sea level, has been damaged by acid rain in the last thirty years or so and of course other forests have suffered as much, but the damage appears to be receding as a result of some more recent developments in technology.

In Britain we have few forests above 1000 metres and not much sunshine in winter. Our prevailing winds come from the Atlantic Ocean and so do not bring us acid rain from other countries. The winds themselves regularly damage our forests as much as or more than acid rain has damaged the forests on the European mainland, but that damage cannot be laid at the door of any human activity.

Our electricity generating stations and factories produce acid fumes and they used to pollute their local surroundings, but the post-war practice of building very tall chimneys largely solved those local problems. When the stations were constructed, it was realised of course that the acids would come down somewhere, but it was believed that little or no damage would be done if they could be spread very widely. So the chimneys were not only very tall; they also had throats at the top to shoot the fumes upward as they emerged, like jets from a hose pipe. Until the Clean Air Act, 1956 effectively eliminated our traditional London smogs, aerial photographs were proudly displayed to show the fumes from the chimneys punching holes through the smog and rising well above it before beginning to disperse. It was not until much later that it was widely realised that the fumes were drifting across the North Sea and polluting lakes and forests in Sweden and other countries.

It should not be difficult to perceive that assessing the value of the damage caused by acid rain from generating stations and ICVs would be just as impossible as assessing the value of the damage caused by exhaust emissions from ICVs in towns. And there would be the additional difficulty that the dispersal of acid rain by wind and weather is erratic; so much so that it would be difficult to ascertain unequivocally from which particular sources the acid rain on a particular lake or forest was coming. And of course to halve the total emission over the whole of Europe would not necessarily halve the damage; it might greatly reduce the damage in many areas or it might have little effect.

Acid rain can be and is being reduced to much lower levels by a combination of the catalytic converters already compulsory in petrol cars, equivalent equipment on diesel cars and lorries, burning natural gas instead of coal and oil for electricity generation and industrial processes, fitting flue gas desulphurisation plant to large furnaces which continue to burn those fuels and by nuclear power. None of those measures will wreck our economy and in any combination determined by the cost method they will not increase the costs of electricity and heat in our homes and work places by more than a few percent – say 10% at most and probably much less. Those two considerations put acid rain, along with oil spillages from supertankers, in the first of our two categories, in which the desired reductions are unlikely to be achieved by trying to make the people and organisations responsible for the emissions pay for the damage they cause, or by taxation. Regulations which gradually reduce the quantities emitted to very low levels are more effective. That is the general approach which has been adopted in the European Community. Since most of this type of pollution comes from electricity generating stations, which are few in number, and from ICVs, which are tested annually for other purposes, the agreed rates of reduction should not be too difficult to monitor and enforce.

GLOBAL WARMING

Global warming may turn out to be by far the worst long-term environmental problem the world has ever faced, but at present it is still chiefly characterised by the high degree of uncertainty attaching to all its aspects.

Apart from varying proportions of water vapour, the earth's atmosphere is composed mainly of nitrogen (80% by volumc) and oxygen (20%) plus very small percentages of other gases, notably 0.035% on average of carbon dioxide. The concentration of carbon dioxide has increased by about 75% in the last 200 years, since the start of the industrial revolution, as vast tonnages of coal and oil (and latterly natural gas), which until then had lain or been trapped under the ground or sea for millions of years, have been exploited and burned.

Carbon dioxide is one of the so-called 'greenhouse' gases which, like the glass in a greenhouse, are transparent to the sun's rays but absorb a proportion of the energy which is reradiated at longer, infrared wavelengths from the earth back into space. The other greenhouse gases are the chlorofluorocarbon gases, nitrous oxide and methane, and they are also on the increase as a result of human activities[6].

The greenhouse effect keeps the earth about 30°C warmer than it would otherwise be and it is essential to life on the planet, but it is feared that, if the concentrations of the greenhouse gases continue to increase at the present rates, the earth's temperature will also increase – very slowly but with some undesirable and perhaps disastrous consequences. The human-made emissions of carbon dioxide are estimated to have contributed about half the global warming during the 1980s and the other three greenhouse gases together made up the other half, but unfortunately it is extremely difficult to predict the future rate of warming because the mechanisms of production, reabsorption and decomposition are complicated and difficult to measure.

To take the apparently simplest example, 75% more carbon dioxide in the atmosphere must enable plants in general to absorb carbon dioxide by photosynthesis faster than they did 200 years ago, but higher temperatures increase the rate of respiration, which produces carbon dioxide, so the net result may be to release as much as or more carbon dioxide into the atmosphere than is absorbed by the faster photosynthesis.

More significantly, the human-made emissions of carbon dioxide are small in comparison with the natural exchanges between the atmosphere and the oceans, and those exchanges are also sensitive to both carbon dioxide concentration and temperature[7]. The temperature also affects the quantities of ice retained in the polar ice sheets and there are other complications.

It follows that to estimate how much the earth's temperature will increase over a long period, it is necessary to make a model of not just the atmosphere but also the oceans, the polar ice sheets, the plant and animal life, and of the earth itself and the mechanisms of emission, exchange, reabsorption and decomposition of all the greenhouse gases. We came across a model in Chapters Four and Six, but that was a simple economic model which had only to include such data as were available or could be collected at a reasonable cost. It did not try to predict what would happen in the distant future, so it could not be faulted for not doing so. Modelling global warming, on the other hand, is for the express purpose of predicting what will happen in the distant future in the way of temperature increases and the consequences of those increases all over the earth. No respected work has made any bones about the uncertainties.

In 1992 the Intergovernmental Panel on Climate Change (IPCC) could 'calculate with confidence' that the human-made emissions of carbon dioxide, chlorofluorocarbons and nitrous oxide, which have lives of 50 to 200 years in the atmosphere before they are reabsorbed or decomposed, would have to be cut by 60% immediately to stabilise their concentrations at today's levels; and that methane, which has a life of only ten years in the atmosphere, would require a 15–20% reduction[8].

There can obviously be no possibility of reductions on those scales, so the report went on to try and forecast the consequences of no reductions (the 'business as usual' scenario) and of some smaller reductions. Its best estimate for the business as usual scenario was that, in the absence of any counter measures, the average temperature might increase by between 0.5 and 1.5°C by the year 2025 and by between 2 and 5°C by the year 2100, with further increases in subsequent centuries, arising from the long lives of the first three gases, even if substantial reductions in human-made emissions then came into effect. Greater increases would probably occur in some regions.

One consequence of such global warming would be a rise in the average sea level. The estimate in the report was that, by the year 2100, it might rise by between 30 and 110 cm (12 to 43 inches), mostly due to the thermal expansion of the oceans and faster melting of mountain glaciers, with only a minor contribution from melting of the polar ice sheets. Some low-lying areas would be flooded and violent storms would probably be more frequent in some regions – less so in others. The higher temperatures would probably displace some plant species to higher latitudes and altitudes and some rare species might be extinguished. The report's main recommendation was for more research to reduce the uncertainties in all the predictions.

That report was criticised for being too timid on one hand and too alarmist on the other. In 1992 the industrial nations agreed to stabilise their emissions of carbon dioxide at the 1990 levels by the end of the decade. Three years later in the Berlin Mandate, a document of fewer than a thousand words, they agreed to draw up within a further two years detailed targets and timetables for cutting their emissions after the year 2000. But some of the leading nations were very reluctant to make either of those undertakings and they did not signal the end of the controversy.

In 1997, at the Kyoto Conference in Japan, the representatives of 39 industrial nations signed a protocol – not a treaty – in which each undertook to 'implement and/or further elaborate policies and measures in accordance with its national circumstances', directed towards a 'quantified emission limitation or reduction commitment' of greenhouse gases. Their collective target was to reduce their emissions by not less than 5% below the 1990 levels by 2012 at the latest. Twenty-seven of them agreed to reduction targets of 8%; the targets of the rest were lower or in three cases limited increases of up to

10%.[9] However the protocol did not include any penalties for failing to meet those targets, and in any case the 5% figure was the outcome of political negotiations, not a scientific estimate of what might be necessary.

Perhaps, as more scientific evidence is gathered, the uncertainties may be reduced and the necessity for the whole world to take the problem really seriously may become obvious. On the other hand it is by no means impossible that global warming will be found to be so slight that there is no need to take any action before the next ice age. Meanwhile it is not a sign of complacency to refrain from taking sides in this dispute; rather it is a sensible refusal to enter a controversy with no new evidence. What remains to be said in principle is that, if a more definite, world-wide commitment comes about, the only practical way of reducing industrial emissions of carbon dioxide will be to tax the human-made emissions from the combustion of coal, oil and natural gas.

To some readers, that statement may sound obvious, but there are many people to whom it is anathema. Even if they are convinced that some strong action is necessary, they will oppose taxation as the means, such is their mistrust of money. Some will argue that surely voluntary reductions by the wealthy nations could achieve a great deal, but do not specify how those nations should apportion the reductions among their industries. Fuller descriptions of the difficulties of dispensing with costs in designing applications of technology are contained in the earlier chapters of this book, particularly Chapter Two.

The same principles must apply to the other greenhouse gases – methane, nitrous oxide, the chlorofluorocarbon gases and some others – but with the added complication that the first two are the products of agriculture – often primitive agriculture – and their sources are correspondingly diffuse and difficult to pin down. It might turn out that the overall reductions of greenhouse gases will have to be predominately of human-made carbon dioxide for no better reason than that its sources are comparatively concentrated and not too difficult to locate.

If taxation is put forward tentatively as the best way forward, some opponents will attack the word 'best' and ask, rhetorically: best for whom? the rich no doubt! By substituting 'only' for 'best', we change the proposition fundamentally. It is now based on a series of three arguments: (1) that the reductions in the rate of carbon dioxide emission will have to be massive – so massive that only the most strenuous efforts by technologists all round the world will be able to bring it about; (2) that those efforts will affect a vast range of projects and (3) that the designers and operators of those projects will continue to use the cost method willy-nilly, unless they are forbidden or physically prevented from doing so, because it is the only rational method and the only method by which they can achieve anything substantial without wrecking the world economy in the process.

One has to refer to taxing carbon dioxide emissions, rather than taxing the combustion of coal, oil and gas according to their carbon content, because it is possible to conceive of removing some of the carbon dioxide from the products of large-scale combustion and disposing of it permanently on the floor of an ocean – or something like that. However that is at best a remote prospect and it will no doubt be more practical to tax the carbon content of those fuels at convenient points between their extraction and eventual combustion – and then allow a rebate to any form of combustion which releases less than the normal quantity of carbon dioxide to the atmosphere.

The vital feature of the taxation method applied to individuals is that it enables them to choose how best to adjust their lives to depend less on the combustion of those fuels. They can put more roof and wall insulation in their houses, invest in more efficient boilers, turn down their thermostats, run smaller cars, walk or cycle to work and so on. Any alternative would inevitably involve some form of fuel rationing, which would run up against the very difficulties we encountered in Chapter Eight in the section on the failure of communism. There is no efficient substitute for allowing people to choose how they will spend their incomes.

More importantly, again as in the same section of Chapter Eight, however ingenious the advocates of rationing may be in dealing with individuals, they have nothing remotely effective to propose for allocating fuel between, say, electricity generation, steel making and aluminium production – or any of the other industries which make up a modern economy. It is in dealing with industry, where the really large economies will have to be made, that the necessity of substituting 'only' for 'best' is forced on the people who instinctively mistrust money.

In general, once a society's concept of prosperity has been defined or redefined and the appropriate taxation regime has been set up, its technologists can start devising the most cost-effective way of moving towards the new aspirations, and it is wise to hamper them as little as possible in specifying how they should go about it, even if it seems obvious to common sense. Common sense has its place but it is sometimes very misleading, as we have seen in the case of electric road vehicles.

Energy conservation

In most, though not all, applications of technology, carbon dioxide emissions can be reduced by a general policy of energy conservation. The well recognised exceptions are electricity generation from nuclear fuels, wind, tides, waves and subterranean heat sources. What is rather often overlooked is that energy conservation is a means to an end, not an end in itself. If in twenty years time, the further evidence now being collected shows or confirms that the earth's carbon dioxide reabsorption mechanisms cannot cope with the additional emissions due to industrial activities without some disastrous long-term effects, energy conservation will have to be pursued with much greater vigour than it is at present, and carbon dioxide taxes will stimulate the individuals and organisations who might otherwise be inclined to drag their feet to give energy conservation the priority it will deserve.

But what if, on the other hand, that new evidence points conclusively in the other direction – towards a confident expectation that those mechanisms will be able to absorb the emissions without any disasters? In that case many of the energy conservation measures which are now being undertaken will prove to have been unnecessary. Not necessarily ill-advised, however, because the various forms of energy will not suddenly become much less valuable in terms of their resource costs and there are few instances as yet of energy being conserved uneconomically in the sense that the discounted present value of the investments in energy conservation, calculated as described in Chapter Three (page 67) without regard to global warming, are negative. Rather they represent the results of closer calculations of the economic benefits. Seen in that light, 'energy conservation' will lose its fashionable appeal as a catch-phrase while remaining a robust means of increasing our prosperity in the long term – perhaps not as fast as some other ways of going about it but with less risk of unforeseen, undesirable consequences.

DISTRIBUTING THE REVENUE FROM TAXATION

As in the case of road pricing to reduce urban traffic congestion, the levels of taxation which are necessary to achieve any desired reduction of environmental pollution or global warming will depend on how much and how quickly people and industries will react to them. Obviously they should be announced as long as possible in advance and slow but inexorable increases are to be preferred to sudden jumps in the rates. Nevertheless some increases may have to be so large as to cause genuine physical hardship to some people and/or impose such a financial burden on some industries as to put them out of business – with disruptive social consequences.

In those circumstances, it will be important not to forget the additional revenue which the new taxes will generate – the higher the taxes the more revenue they will produce. But they will not have been imposed for the purpose of raising revenue, so the hardship/burden may be lessened in some cases by returning the revenue to the taxpayers, not as individuals or individual companies since that would nullify the inducements to reduce their fuel consumption, but to particular sectors and groups which are hard hit by a fuel tax, rather than by reducing general taxes. For example, if steel making were particularly hard hit, the rate of corporation tax payable by steel makers might be reduced – and similarly for any other industry. They would then have the inducement of paying less fuel tax without losing the benefit of having to pay less corporation tax. By adjusting the rates of fuel tax and corporation tax, the firms which did not economise on fuel could be made to subsidise the firms which did.

By the same token people in cold climates might be given compensating tax concessions, which would not reduce the rewards for fuel saving but would cushion those who genuinely could not save fuel. If the concessions were generous at first but then diminished as the savings achieved by the other people reduced the revenue, everybody would be given the fairest opportunity to adjust their life style or industrial process to the new consensus of what constitutes genuine prosperity in the long term.

SUMMARY

In the course of this chapter, in conjunction with the previous chapters, we have evolved, not a solution to any pollution problem but a strategy for using technology effectively to tackle environmental pollution in its various forms. It has nine components, as follow:

1) Classify the cause of the pollution. Is it local, regional or global or a genuine mixture of more than one of those? Resist the temptation to blur the distinction.
2) If the cause is local or mainly local, investigate how it can be eliminated or reduced locally. Do not wait for regional or global agreement on how to go about it unless there is strong, positive evidence that such agreement may soon be achieved. Similarly, if the cause is regional, do not wait patiently for global agreement.
3) On the other hand, if the cause is regional do not try to

deal with it locally, and if it is global do not try to deal with it locally or regionally, unless in either case there is strong, positive evidence that local or regional actions will lead to the other parts of the region or globe, as the case may be, following suit in the very near future.

4) Analyse the problem to determine whether the pollution could be eliminated or substantially reduced by commercially available technology at moderate cost by passing and enforcing non-draconian, physical regulations. If not, but there seems to be a real prospect of such technology becoming available, finance or otherwise encourage the development of that technology, but do not make extensive plans for its adoption until it is genuinely on the verge of commercial availability. Study the history of that technology before making definite plans.
5) If the pollution cannot be eliminated or sufficiently reduced by physical regulations without severely damaging the economy of the society which is suffering from the pollution, extend and deepen the analysis to determine how best it can be gradually reduced. Recognise that the best means may have more than one component.
6) Expect the technologists concerned to employ the cost method. Try not to guess how they will do so in advance. Leave them the maximum scope for exercising their skills and ingenuity. Challenge anyone who opposes the use of the cost method to describe in detail how they would decide on the design and operation of an application of technology without invoking costs.
7) Always consider physical effects which can be measured objectively and costs which can be objectively estimated, rather than appeals to ill-understood concepts, such as energy conservation (without distinguishing between the different kinds of energy) or noise (without taking account of the adaptability of the human ear) and so on.
8) Conduct all analyses openly. Oppose all arguments that information should be withheld on the ground that the general public are not clever enough or too poorly educated to understand it.

9) Recognise that:
i) nothing much can be achieved in a democratic society without public support – or at least public assent,
ii) undemocratic societies have always caused far worse pollution than democratic societies in comparable circumstances,
iii) public opinion can change (think of drink-driving and the wearing of seat belts in cars) but only slowly.

The foregoing list is not intended as a rigid scheme to be followed in the stated order of the items. None of the items is original but some are controversial. If the reader is not satisfied that the pollution with which he or she is particularly concerned is being tackled with reasonably adequate rapidity, he or she may benefit from considering which of the components is being or has been omitted and directing his or her protests accordingly.

REFERENCES AND FOOTNOTES

1) Much of the information in this section came from 'How to seal a supertanker', by Michael Cross and Mick Harmer in the 14 March 1992 issue of *New Scientist*. However the conclusions are the author's own.
2) A prospective new method, mentioned in the *Scientific American* of April 1993, connects the anchor chain to the ship via a windlass, driven by a special hydraulic motor which keeps a constant tension on the chain and so enables the windlass to absorb the energy of the ship as it swings about, without breaking the anchor flukes.
3) European Foundation for the Improvement of Living and Working Conditions 1983. Report no. EF/83/26/EN: 'A European study of commuting and its consequences'.
4) The numerical data and analysis in this section were set out in the COST 302 report – 'Technical and economic conditions for the use of electric road vehicles', which was cited in Chapter Three (Footnote 2)
5) Zinc/bromine, nickel/iron, nickel/zinc and solid-state lithium/nickel batteries were considered in detail in the European study

(in the appendices to Chapter Five) but none of them has become commercially available for road vehicles in the ensuing thirteen years.

6) The chlorofluorocarbon gases were invented in the 1930s and have been used as aerosol propellants, solvents, refrigerants and foam blowing agents. Nitrous oxide is presumed to have arisen from human activities, probably agriculture. Methane comes from rice production, cattle rearing, burning plant matter (known as 'biomass'), coal mining and ventilation of natural gas.
7) 'Oceans and the global carbon cycle' – an introduction to the Biogeochemical Ocean Flux Study of NERC Marine Sciences directorate, Plymouth May 1989.
8) 'CLIMATE CHANGE The IPCC Scientific Assessment', the final report of Working Group 1. The IPCC is jointly sponsored by the World Meteorological Organisation and the United Nations environmental Programme. Two more working groups were set up, one to assess the environmental impacts and socioeconomic consequences of climate change and the other to formulate response strategies, but their reports had not been published at the time of writing.
9) 'Kyoto Protocol to the United Nations Framework Convention on Climate Change', Agenda item 5, December 1997.

10

TECHNOLOGY AND PHILOSOPHY

> The traditional disputes of philosophers are, for the most part, as unwarranted as they are unfruitful. The surest way to end them is to establish beyond question what should be the purpose and method of a philosophical enquiry. And this is by no means so difficult a task as the history of philosophy would lead one to suppose. For if there are any questions which science leaves it to philosophy to answer, a straightforward process of elimination must lead to their discovery.[1]

> We all have our philosophies, whether or not we are aware of the fact, and our philosophies are not worth very much. But the impact of our philosophies upon our actions and our lives is often devastating. This makes it necessary to try to improve our philosophies by criticism. That is the only apology for the continued existence of philosophy which I am able to offer.[2]

It was observed in Chapter One (page 4) that, such is the sheer volume of knowledge in the modern world, our biggest problems are inevitably approached with an inadequate individual understanding. For myself, I could wish that I knew more about the subjects of all the chapters, and it is only because I believe the problems encountered by societies in applying technology are linked together and that the solutions cannot be perceived if the subjects are treated separately, that I have had the temerity to tackle the whole gamut in a single book.

I feel my personal knowledge to be least in the field of philosophy, and I nearly omitted this chapter for that reason. But I have come to realise, and shall try to show now, that the link between technology and philosophy is as important as the other links, and that the neglect of that link by technologists and others is a potential cause of poverty and sometimes tragedy. I do not claim to be the first to discover that link. When it is plainly stated, it appears to be obvious, but it is evidently not so thoroughly understood that it need not be repeated.

Like many technologists (I suspect), I decided early in life that Philosophy – by which I mean the serious study of the works of famous philosophers – was not for me. The philosophy books I dipped into seemed to be excessively and unnecessarily obscure. Then I chanced on A. J. Ayer's *Language, Truth and Logic* (many years after it was published), which I thought I understood. Like many young readers, not yet bowed down and confused by life's burdens and compromises, I was delighted with the passage in my first quotation above, and chuckled when, in his very first chapter, Ayer dismissed metaphysics as a mixture of humdrum errors by philosophers who have been duped by grammar (or have otherwise failed to understand the workings of our language) and literally senseless utterances by mystics. In his introduction to the second edition he regretted that he had written with more passion than most philosophers allow themselves to show, but still believed that the point of view he had expressed was substantially correct. I was therefore heartened by his concluding passage, in which he wrote:

> ...we distinguish between the activity of formulating hypotheses and the activity of displaying the logical relationship of these hypotheses and defining the symbols which occur in them. It is of no importance whether we call one who is engaged in the latter activity a philosopher or a scientist. What we must recognise is that it is necessary for a philosopher to become a scientist, in this sense, if he is to make any substantial contribution towards the growth of human knowledge.

The thought that philosophers should become scientists must have been startling in philosophy circles in the 1930s. However by science Ayer evidently meant pure science, which has no purpose except to accumulate knowledge, regardless of its utility. He gave no hint of a link between philosophy and applied science or technology. It was Sir Karl Popper's remark, 36 years later, in my second quotation that brought home to me that technologists cannot ignore philosophy with impunity. Not that even he – or indeed any of the others as far as I could see – paid any heed to technology, but he hit the nail on the head when he wrote, albeit apologetically, that our philosophies (in the sense of our sets of beliefs) can have devastating effects on our lives, especially if we have not tried to clarify the content of our philosophies.

Indeed we do well to consider, not only the effects of our beliefs on ourselves but also the potential effects of the conflicts between our beliefs and those of the people with whom we deal, especially when their traditions are very different from ours. I was aware that other respected philosophers were inclined to regard Ayer as an *enfant terrible*, but that stigma was not attached to Popper.

In the next two sections of this chapter we shall try to establish what we in the Western world – and particularly technologists in the Western world – actually believe about freewill, personal responsibility and forethought, and then trace some consequences of our beliefs. By 'actually believe' I mean what we have at the back of our minds when we make decisions which we discern will affect our lives – our daily decisions about such things as what to eat and wear and whether to take an umbrella when we go out and our bigger decisions about such things as whether to buy a house or a car or go on a long journey or get married. And in the case of technologists we shall be investigating our (or their) decisions on the design, construction and operation of our (or their) projects. Those actual beliefs are not always the same as what we sometimes declare that we believe in our religious services and philosophical discussions, when such decisions are not at the forefront of our minds – which is not to imply that we are necessarily conscious hypocrites but that we do not always insist that our actual beliefs are totally consistent with our declared beliefs. Such inconsistencies may seem unimportant and very often they are, but sometimes they matter in the sense that they affect our ability to prosper – and help other societies to prosper – from technology, and it is then worth our while at least to acknowledge them and perhaps try to reconcile them.

FREEWILL AND PERSONAL RESPONSIBILITY

> One cannot base one's conduct on the idea that everything is determined, because one does not know what has been determined. Instead one has to adopt the effective theory that one has a free will and that one is responsible for one's actions.[3]

A central preoccupation of philosophers, priests and theologians throughout history and what is known of prehistory has been the linked problems of freewill and personal responsibility. The gods in the old polytheistic religions were portrayed as capricious and liable to have disputes among themselves, which meant that their worshippers could not forecast, even approximately, what the gods might decide to order. That inability left them in a state of perpetual unease and sometimes terror, which could only be partially alleviated by prayer and propitiation.

When polytheism was superseded by monotheism, notably in Judaism and followed by Christianity and Islam, religious leaders taught that obedience to God's known commandments will lead to less suffering on earth – or at least the fortitude to bear it – and eventual contentment in the after-life. That philosophy implies that people have free wills to choose whether to obey the commandments or not and also, since the commandments do not cover every eventuality in detail, to judge how to apply them in the variety of situations with which people have to contend.

But all the monotheistic religions have retained at least an element of fatalism. Pure fatalism is the doctrine that all events happen by unavoidable necessity, but as far as I am aware no theologian would expect anyone to believe anything so absurd as that. Nevertheless believers are routinely enjoined to give thanks to God when things go well (however much they may secretly believe that their success was the outcome of their own efforts) and to submit humbly to His will when they seem to go badly wrong.

In the Western world, modern science and technology have, by and large, reduced the fatalistic element in our beliefs and increased the scope for people to exercise freewill and succeed – or fail – by their own and their friends' and associates' efforts. But there has been one extremely damaging exception, namely the universal determinism of Hobbes and Laplace – the enormous theory that there is a scheme of strict causality regulating the sequence of all phenomena, down to the individual atoms, so that the whole future of everything in the universe is predetermined and could theoretically be predicted in detail, though never (if you read the small print) by any human being. It is difficult to see how a genuine believer in universal determinism can logically avoid becoming a total fatalist.

However loosely or tightly fatalism and universal determinism may be linked in academic philosophical, theological and pure-science circles, they are both instinctively rejected as nonsense, or simply ignored, by most other people in the Western world – certainly by Western technologists. We believe in causality, but our most careful applications of the laws of physics fail to give us predictable results far too often for us to believe in universal causality in anything approaching the Laplace sense. The unpredictability of the turbulent flow of liquids and gases in pipes is a good enough example. It has interfered in the design of electricity generating stations, chemical process plants and most other projects to some extent throughout their history. The recent Theory of Chaos, though not yet completely accepted by the physics establishment, is putting paid to the possibility of a deterministic solution of a wide range of important problems in applied science.

At the same time, our whole lives are based on our confident belief in our own freewill. So are the lives of philosophers, theologians and theoretical physicists whenever they turn aside from their work to eat, drink and take their recreation, but in their working lives theoretical physicists were apparently in the intellectual thrall of universal determinism from the time of Laplace until about seventy years ago[4]. Only then did they begin to wrestle with that dragon in their learned discourses and they have been doing so from time to time ever since. The essay by Stephen Hawking, the world-famous physicist and best-selling author, from which I took the above quotation, shows that the dragon still sometimes lifts its ugly head in the highest pure-science circles, but they seem to be getting the better of it at last.

The unsupported instincts of technologists and others can be unreliable and they are normally looked at askance by philosophers. Fortunately in this case instinct can be supported by a logical argument which has recently attained philosophical respectability. We begin by averring that, even without bringing in such phenomena as the turbulence of flowing fluids, fatalism and universal determinism are contrary to our common sense.

To take just one example among many, if we have been arguing for years about the route of a motorway, and have finally built it along just one of half a dozen proposed routes, all fiercely advocated, attacked and defended by different parties, it would be absurd (to say nothing of it being irreverent) to believe that God or Fate had decreed from the outset that it should follow that particular route, or that it was the predetermined outcome of the laws of physics. And of course it would be equally absurd to suppose that the millions of decisions taken in the design of an electricity generating station, or any other application of technology, were predestined in any such manner.

If any members of a project design team started to believe that their decisions were preordained or predetermined, they would jeopardise the whole project – unless their dangerous madness was detected and they were reformed or expelled.

And if the sufferers from a horrific disaster in a Western industrial country can be comforted by their belief in God, the technologists who were concerned with the project will nevertheless be censured and penalised if the enquiry shows that they were seriously to blame. They will not escape by pleading that the disaster was preordained. Determinism and fatalism do not devastate our lives in the successful Western industrial countries for the very reason that they are not part of what the people in those countries actually believe, as distinct from what some of them discuss when nothing of any personal consequence to themselves depends on their conclusions.

Some critics may enter the discussion at this point to assert that common sense is no more reliable than instinct as a basis for a philosophical argument and that we cannot know for certain that God or Fate or the laws of physics have not decreed or predetermined everything which happens. Until a couple of decades ago those critics might have been taken seriously in philosophical circles, but nowadays they are out of touch with contemporary academic thought. After the remark which I have quoted, Popper, a world-famous philosopher, continued immediately to write: 'Science, philosophy, rational thought, must all start from common sense.' Not, in his view, because common sense is a secure starting point but 'because we do not aim to try to build (as did Descartes or Spinoza or Locke or Berkeley or Kant) a secure system . . . Thus we begin with a vague starting-point, and we build on insecure foundations. But we can make progress . . .'

Thus has at least one leading school of philosophy begun to say things which non-philosophers can understand. We should be grateful to Sir Karl for distancing modern philosophy from no fewer than five illustrious prior philosophical schools, since only someone of his standing could do so with the necessary authority. At the same time we can legitimately remind his followers that ordinary people – not least technologists – are generally more involved in making decisions with immediate or short-term impacts than most academic philosophers. And it is those impacts which build up our fund of common sense, on which we all – philosophers and non-philosophers alike – rely in our daily lives. So perhaps those philosophers should try sometimes to understand and respect what we have to say, rather than always leaving us to trail after them. We may also wonder if they realise how wide is the gate which their leader opened. Meanwhile we can accept that, as Popper went on to say, we should not rely on common sense to make that progress.

Indeed it is evident that progress in both philosophy and technology has come from challenging common sense and showing that it is often wrong or that a different belief is necessary or desirable. For two examples, the common sense belief that the earth is flat (Popper's example) was rejected and replaced by the powerful theory – which is essential for ocean navigation – that it is a globe, and more recently the common sense belief that bleeding some steam from a turbine would be wasteful was rejected and replaced by the valuable theory that it can make the turbine more efficient, as described in Chapter Two (page 32), and so, up to a point, more economical.

After a time the new theories are widely absorbed and become part of our common sense. Thus only a few cranks still believe the earth is flat and no technologist of repute would deny the efficacy of bled-steam boiler feedwater heating. This book has some new challenges to present tenets of common sense, and I naturally hope that they will eventually be absorbed in the same way. But in every case, the onus of proof has rested and must always rest with the challenger. That is the only way of preserving our sanity and our civilisation.

And that is precisely how the theory of universal determinism has failed. Its defenders, if there are still any, have failed to show that our common sense view is wrong or that the theory is necessary or desirable. It does not add to our knowledge and it does not enable us to do anything better than the common sense view. So whilst it will be a relief when those defenders finally admit that they have rejected their pet theory, we need no longer seriously care how soon or late they come to do so.

We may seem to be reaching the position where technologists try to dictate to non-technologists which philosophy they must adopt – that they must believe in freewill and personal responsibility. In truth we are not so arrogant as that, but we should be hypocrites if we denied that a society which aspires to prosper from technology will be disappointed if it does not espouse both those beliefs. If you come across a qualified technologist who is inclined to deny that proposition or prevaricate on the subject, just ask him or her how he or she would organise a team of designers and/or manufacturers and/or builders and/or operators of a project if, whenever something goes wrong, they all – or even a substantial proportion of them – shrug their shoulders and recite a pious phrase about Fate or the will of God or the equivalent in their religion, or start talking about universal determinism.

How have we arrived at this point? It seems plausible to suppose that, when the human race started to think about the future but still had only a few primitive tools and was in continuous danger of extinction from wild beasts, storms, floods, crop failures and diseases, a high degree of fatalism was an inevitable part of any set of beliefs. People could not otherwise reconcile themselves to the misery in which thinking about the future would otherwise have drowned them. One would have to study the history of theology to know whether that supposition is supported or contradicted by the evidence. However that may be, our present confidence that we can choose what we can do next, and that we can often – though not always – increase our material prosperity by making good choices, comes from our modern technology, which is of course derived from modern science. And that confidence did not start to become widespread until about a hundred and fifty years ago, even in the West.

For example, until about the middle of the nineteenth century, a long sea voyage was still hazardous. You were wise not to set sail in the winter and when you did depart you took with you the blessings and fervent prayers of your friends and relatives. Nowadays you can pop across to New York by air for a business trip and be back almost before your absence has been noticed. You may, of course, have half a dozen business and family reasons for mentioning your intention in advance but, if we exclude those, you are as likely to be concerned with making sure that someone will feed your cat as seriously worried about any danger to yourself. Accidents still do happen, but the risk is not much greater than the risk of staying at home, where you may be involved in a road accident when you go to the supermarket. If you perish, your relatives may pray for your soul, but they will also apply for a large monetary recompense from the airline – as the aircraft designers are well aware when they come to decide how much they dare cut down the weight of the plane to save fuel.

Aesthetic and moral objections

We in the West have reached that comfortable situation gradually. Technological progress and faith in our own collective abilities have gone hand in hand over the centuries, slowly at first but at an ever increasing pace. Some literary intellectuals have professed to be aesthetically revolted and some religious leaders and other moralists have wrung their hands and begged us to beware of the spiritual dangers of materialism, but they have largely been ignored. C. P. Snow (later Lord Snow), who had the rare distinction of being both a famous literary intellectual and a scientist and whom I quoted in Chapter One (Footnote 3), dealt plainly with the aesthetic objection. He wrote:

> ...one truth is straightforward. Industrialisation is the only hope of the poor. I use the word 'hope' in a crude and prosaic sense. I have not much use for the moral sensibility of anyone who is too refined to use it so. It is all very well for us, sitting pretty, to think that material standards of living don't matter all that much. It is all very well for one, as a personal choice, to reject industrialisation – do a modern Walden, if you like, and if you go without much food, see most of your children die in infancy, despise the comforts of literacy, accept twenty years off your own life, then I respect you for the strength of your aesthetic revulsion. But I don't respect

> you in the slightest if, even passively, you try to impose the same choice on others who are not free to choose. In fact, we know what their choice would be. For, with singular unanimity, in any country where they have had the chance, the poor have walked off the land into the factories as fast as the factories could take them.
>
> ... It was no fun being an agricultural labourer in the mid to late eighteenth century, in the time when we, snobs that we are, think of only as the time of the enlightenment and Jane Austen.
>
> The industrial revolution looked very different according to whether one saw it from above or below. It looks very different today according to whether one sees it from Chelsea or from a village in Asia...

Turning to the spiritual dangers, in Chapter One (page 4), we sidestepped the questions whether prosperity brings happiness and whether such things as family cohesion, social stability and national defence are less important than prosperity. We postponed them but now we can give, if not answers at least some pertinent comments, by referring to Popper's proposition that science, philosophy and rational thought must all start from common sense. Our common sense tells us that, although prosperity cannot guarantee happiness, there is no compelling reason in our society why it should inevitably stop us from being happy. Some prosperous people are happy and others unhappy, and the same is true of poor people, but we suspect the proportion of unhappy people is larger among the poor than among the prosperous.

We can watch with some amusement the antics of people who claim to believe in 'the good life' without the so-called 'trappings of modern civilisation', but who retain an array of technically sophisticated implements and services which would have made their forebears a couple of centuries ago blink with envy. More seriously, perhaps if you have been told repeatedly as a child in a poor family that it is wicked to be rich, then even if you do not altogether believe it you may find that that creed interferes with your enjoyment of prosperity in adult life. Or perhaps, if you have had little or no education, you may behave so foolishly when you become prosperous as to make you unhappy. Perhaps it is easier to be happy if the other people you meet are not too excessively richer or poorer than you.

Those are just three out of scores of possible hypotheses, none of which invalidates our common-sense assumption that the prosperity technologists bring to our community will probably bring some happiness in its train, unless and until it can be demonstrated that it does not. And by the same line of argument we can assume that prosperity in itself is not likely to weaken our social cohesion and that the more prosperous we are the more easily we shall be able to defend ourselves. How much of our prosperity we should give up to prepare for a possible military emergency will depend on our assessment of the risks, which will depend in turn on the available evidence at the time.

FORETHOUGHT

The other belief which has gone hand in hand with our technological progress and our increasing faith in our collective abilities is our belief in the value of forethought. It is as important as freewill and personal responsibility; indeed it is difficult to imagine any of those three qualities without the other two. For some unexplained reason forethought is not much discussed among academic philosophers, but it is none the less important for that.

By forethought and good technology, we can be warm and well fed throughout most winters and we can be much healthier than our forebears. We can enjoy a range of comforts and activities which were unknown to them. We cannot control the weather but we can forecast it a few days ahead and so avoid being exposed to its worst rigours while continuing our pursuits when it is clement. We cannot forecast when earthquakes will occur but we are gradually learning how to erect buildings which will stand up through earthquakes and so reduce the loss of life from thousands to dozens and damage to property in the same proportion.

However, the habit of forethought seems to be more difficult to teach than belief in freewill and personal responsibility. We have largely escaped from the thrall of fatalism in the West and, although individually we may try to escape from our responsibilities and we succeed more often than is good for our community, we can be taught that we ought not to do so. But millions of people in the most advanced industrial economies go from cradle to grave without acquiring the habit of forethought beyond a few days. If they do not live precisely from hand to mouth, they handle their finances from week to week and cling tenaciously to that way of life. They are consequently much less prosperous than they could be, even though they are cosseted by special arrangements for doling out their wages and other entitlements at weekly intervals. Since they are apparently impervious to education in that subject they have to be tolerated. They are less of an asset than they might be in a society which must routinely plan nowadays for two or three decades ahead and in some respects for a hundred years ahead. That is not because they are unskilled (many have valuable skills) but because of their mental myopia.

Perhaps one reason behind this educational problem is the Christian teaching: 'Take therefore no thought for the morrow: for the morrow shall take thought for the things of itself. Sufficient unto the day is the evil thereof.'[5] At the risk of offending some sincere Christians, one has to point out that a society whose members actually accept and act on that teaching in their everyday lives, as distinct from just listening to it in church from time to time, will never prosper from technology. One of the reasons why we in the West have prospered and do prosper from technology is that we consistently do take thought for the morrow. Perhaps we should do better to cease teaching our children to recite what we do not ourselves believe.

OTHER PHILOSOPHIES AND TRADITIONS

Modern science and technology have been created almost entirely by men and women who were brought up in the

Christian or Jewish faith. They have not by any means all been 'good' Christians or 'good' Jews, indeed many of them have either lapsed from or explicitly rejected those faiths, but they have retained their beliefs in freewill, the importance of forethought and the principle of duty.

The question which presents itself in that context is: can a society whose members are not imbued with those beliefs prosper from modern technology? It is a hypothetical question in that it does not formally assume that there are any such societies, but there are grounds, surely, for thinking that there may be. Let us start with a fictional example.

Neville Shute was a technologist (or engineer if you prefer) and a best-selling novelist. In *A Town Like Alice*, he recounted how a young English woman was allowed to live with her companions and a child for three years in a Malayan village, where they passed as native workers in the rice fields during the Japanese occupation in the Second World War. When the war was over she returned to England and inherited some money. That inheritance enabled her to go back to the Malayan village and present the women who had befriended her with a well, thus releasing them from the burden of carrying drinking water for the village nearly a mile every day.

The question is: how much benefit was the well supposed to bring to the women and the village? In the short term the women would have an easier life, but the picture one has of such societies (which the novel depicted as Muslim) is that they are – or were then – completely male dominated and that the men concerned themselves mainly with soldiering, local politics and fathering children. They did not engage in agriculture or otherwise contribute to the prosperity of the village.

In that case the greater well-being of those women might have been short-lived. On returning from the war, their menfolk might have filled their initially easier lives with more child bearing and work in the fields to raise crops for the larger population. The need of that larger population for drinking water could be met by drawing supplies from a deeper level than was previously possible. If drought was one of the occasional scourges which limited the population, it would do so at longer intervals because of the greater depth from which water could now be drawn, but eventually even the new well would run dry and the larger population would suffer more than the smaller one would have done before the well was dug.

That episode and the supposed sequel are both fictional and so might be dismissed as having no basis in fact. Some settlements in the Sahara Desert suffered in the manner described after digging deep wells, financed by foreign aid, but Shute's village was depicted as being in a fertile region. Very well, let us turn to a factual occurrence.

> In December 1984, Bhopäl [city and capital of Madhya Pradesh in central India] was the site of the worst industrial accident in history, when about 45 tons of the dangerous gas methyl isocyanate escaped from an insecticide plant in Bhopäl ... owned by the Indian subsidiary of the U.S. firm Union Carbide Corp. The gas drifted over the densely populated neighbourhoods around the plant, killing many of the inhabitants immediately and creating a panic as tens of thousands of others attempted to flee the city. The final death toll was estimated at as high as 2,500 lives, and local medical facilities were overwhelmed by about 50,000 other people who were temporarily disabled by respiratory problems and eye irritation resulting from exposure to the toxic gas. Investigations later established that substandard operating and safety procedures ... had led to the catastrophe.[6]

I have not studied the investigations to judge whether the operating staff, or any of them, were imbued with any degree of fatalism. It would take months, if not years, of time and effort to do so, with no certainty of coming to a firm conclusion which would convince anyone who was predisposed to take the opposite view. The relevance of the catastrophe in the present context is not in who was to blame but to postulate that, if some of the staff had based their working actions on a belief in fatalism, or if they were unconsciously influenced by such a belief, a frightful accident of some kind sooner or later would have been a probability rather than a remote possibility.

There may be clever philosophical or theological arguments for accepting a degree of fatalism in one's life while retaining a sense of responsibility for one's actions. But the Bhopäl plant was not staffed by philosophers or theologians; it was staffed by ordinary folk and it was their understanding of what they had been taught or had otherwise absorbed in childhood that was relevant.

Intermediate Technology

That insecticide plant was very large and it used sophisticated processes far beyond the understanding of the local population. It has been postulated that the root cause of the disaster was that the technology was inappropriate, that what is needed is so-called 'Intermediate Technology', which a local population can learn and apply themselves, using mainly locally available materials.

For example Nepal in India is said to have the potential to produce all its power in the form of 'micro-hydro' power from its rainfall. Yet women in Nepal may spend five hours grinding a single day's corn by hand. The turbines for such micro-hydro schemes can, it is said, be made in rural areas by local artisans using existing metal working techniques.

And in Kenya, it is said, the women commonly go on ever longer treks for increasingly scarce fuel wood, which they then burn, inefficiently and unhealthily, on open 'three-stone' fires. They can be taught to use simple clay 'Upesi' stoves, which need less wood and emit less smoke, so their health improves. It is said that the stoves can be made locally, using existing pottery skills, at a cost of less than £1 each.

The typically Western response to such efforts to help poor people in remote places was well expressed by a bishop, who said:

> I believe that Intermediate Technology offers one of the most brilliantly effective ways of helping poor people in developing countries to transform their own life. I have

> seen the amazing difference that this kind of assistance can give and there would be few more obviously fruitful methods of encouraging true development than supporting this exciting and creative work.[7]

But are there not some unanswered philosophical questions? Why have the inhabitants of those regions not invented hydropower and efficient clay stoves themselves, many years ago? Water mills were in common use in Western Europe in the eighteenth century. Was not their invention largely the outcome of Western European beliefs? The name, 'Intermediate Technology' implies that Western technology can be introduced in stages, such as 'elementary', 'intermediate' and 'advanced'. But can a population with very different beliefs and traditions follow such a route without disrupting its social cohesion?

Some people will argue that such doubts should not hold anyone back, that they must go forward in faith to do the Will of God as they understand it. But some technologists, including myself, keep remembering that that kind of impulse to help people in remote places, when it has guided the introduction of technology in the past, has had often damaging and sometimes disastrous consequences. Intermediate Technology will not lead to another Bhopäl, but it may have other consequences which have not been foreseen.

An academic philosopher might challenge the bishop to explain just what he meant by 'true development'. Did he mean development towards greater material prosperity, with its moral dangers, or towards a more Christian society? And if the latter, in what way is the Christian religion superior or preferable to the religions which have prevailed in those regions for centuries? Can he be sure that intermediate technology will not disrupt the moral fabric of those societies without even bringing any long term prosperity to them?

The common sense view is that material prosperity is the aim of technology and that we have no evidence that it does not increase our happiness, social cohesion and ability to defend our way of life – but that is in our society with our ingrained beliefs in freewill, personal responsibility and the value of forethought. It may be that intermediate technology can bring the same benefits in Nepal, Kenya and other remote regions, but that can hardly be asserted as a matter of common sense, now that we have thought about their and our philosophical beliefs

So the discussion runs eventually into the sand, as predicted in Chapter One, but perhaps not for quite the same reason as philosophical discussions commonly run into the sand. If it can be firmly concluded that belief in freewill, personal responsibility and the importance of forethought are essential prerequisites for a society to prosper from technology, that surely is a valuable achievement. If you do not think the arguments in this chapter lead to any such conclusion, you may at least be stimulated to think out your reasons why they do not.

If you do accept that conclusion, you may go on to ask: what is to be done about it ? That is the instinctive Western reaction to almost any new problem – to demand that somebody (usually the government) shall take steps to solve the problem, or at least mitigate it. If people in underdeveloped countries are misapplying technology to their own disadvantage because of their philosophical beliefs, the argument runs that we must set about showing them the way by re-educating them. But how can people with one set of beliefs systematically re-educate people with a radically different set of beliefs ? My own impression is that one set of beliefs may invade and eventually strangle or otherwise supersede another set, but not systematically or in any organised manner. So that discussion also runs into the sand for the present, which is not to say that *research* in the subject should be abandoned – merely that action should follow research, not try to go in front.

REFERENCES AND FOOTNOTES

1) A. J. Ayer, *Language, Truth and Logic*, Victor Gollancz, 1936.
2) Sir Karl Popper, *Objective Knowledge: An Evolutionary Approach*, 1972, by permission of Oxford University Press.
3) Stephen Hawking, 'Is Everything Determined' in *Black Holes and Baby Universes and other essays*, Bantam Press, 1993.
4) For example, Sir Arthur Eddington, the world-famous theoretical physicist of his day, wrote about the decline of universal determinism in at least three books, *The Nature of the Physical World* in 1928, *New Pathways in Science* in 1935 and *The Philosophy of Physical Science* in 1939. In *New Pathways*, he wrote: 'Ten years ago [i.e. in 1925] practically every physicist believed himself to be a [universal] determinist ... Then rather suddenly determinism faded out of theoretical physics ... Some writers are incredulous ... Some decide cynically to wait and see if determinism fades in again.'
5) The Gospel According to St Matthew, Chapter 6, verse 24.
6) From "Bhopal" in *Encyclopaedia Britannica*, 15th edition (1992), 2: 189
7) The Right Reverend Simon Barrington Wood, Bishop of Coventry.

11

CONCLUSIONS

The time has come to pull together the threads of the arguments in the foregoing chapters, as promised in Chapter One (page 8), and summarise the conclusions.

Modern technology is derived from the modern physical sciences. They constitute a comprehensive, highly organised body of knowledge. But an application of technology cannot be designed simply from the laws and principles of the physical sciences. There is always a wide range of designs, any of which would comply with those laws and principles. To derive prosperity from its technology, a society must allow and ensure that the designers explicitly relate their decisions to the costs of the resources which are employed in it.

Costs and prices are commonly regarded, consciously or unconsciously, as sordid nuisances which constrict and impoverish people's lives. They may be granted the status of a necessary evil, but an evil nevertheless which one would like to circumvent. Technologists sometimes slip into that way of thinking, but the truth is that, without the complex web of costs and prices which is to be found in any successful industrial society, its technologists would not know how to proceed and their projects would be as likely to impoverish the community as enrich it. The concept of energy, although abstract, is useful up to a point, but it is misleading when substituted for costs and prices, which are the lifeblood of technology in peacetime.

When that proposition is put thus plainly, it is still too easy to assent to it, but then to ignore its consequences, which are far reaching. For a successful outcome, the cost method, which was explained in Chapter Three, must be consciously and conscientiously applied. It requires careful attention to how the cost estimates are obtained, when the payments are to be made, the effects of monetary inflation, the difference between commercial and resource costs (including the effects of taxes and subsidies), how best to cater for uncertainties and how to organise the necessary co-operation between the people engaged in the project. The only primary sources of data for the cost method are the accounts of the commercial transactions of the organisations concerned.

Large-scale applications of technology involve large, long-term investments in tangible fixed assets. In as much as designers habitually think in terms of the whole period of manufacture, construction, operation and eventual disposal of those assets, they are not much concerned with or affected by the conventions adopted in compiling published annual accounts, although they are of course concerned with the internal management accounts. However, technologists understand the two main causes of depreciation, by which the foreseeable deprival value of a tangible fixed asset declines, year by year, from its initial acquisition cost to its final disposal value. Those causes are physical deterioration, the minor cause, and technological progress, the major cause. They are the basis of a theory of depreciation, developed in Chapter Four, with important implications.

It emerges from that understanding and theory that the convention most frequently adopted for the depreciation of tangible fixed assets in annual accounts, the so-called straight-line method, seriously overvalues the assets throughout their economic lives and so undermines the usefulness of those accounts. They overstate the profitability of a company or undertaking when it is expanding and/or understate the prices which should be charged to balance its revenue and its expenditure. Consequently, when the expansion ends (which may happen unexpectedly) and is followed by contraction, the difficulties of adjustment are exacerbated. By substituting a formula which is as simple arithmetically as the straight-line formula, the theory enables the annual depreciation and the prices and profits of the organisation to be calculated more realistically and the difficulties of adjusting to an expansion/contraction reversal reduced.

It also emerges that Current Cost Accounting, which was introduced (but later discarded) to counter the distorting effects of monetary inflation on Historical Cost accounts, is based on a technological fallacy. A much simpler method of adjusting Historical Cost accounts for inflation would remove those distortions, which in the absence of any adjustment overstate the profitability of companies and undertakings in much the same way as the straight-line convention for calculating depreciation.

The cost method is a powerful tool in the hands of designers of technological projects. It builds up industries in some places and reduces them – sometimes destroys them – in others. It is a necessary, but not a sufficient condition for deriving prosperity from technology. In general, the necessary concomitant is a set of rational prices and methods of charging for goods and services across the whole economy. By an extension of the theory of depreciation and using Adjusted Historical Cost Accounting for inflation, a method of charging and calculating such prices has been outlined for public electricity supply, as an example of a commercial service in the public sector. The difficulties in that case are peculiarly severe because electricity cannot be stored (except in insignificant quantities), but they are not insurmountable. That example points the way to devising rational prices and methods of charging for other commercial services in the public sector, as a precursor to the regulation of such services in the private sector.

Some of the commercial public services in a modern industrial society, notably the so-called utilities, are natural monopolies or oligopolies. In some cases it is physically impracticable to provide more than one – or a very few – sources of such services for any customer or consumer. In other cases, the practicable number is severely restricted by the economies of large-scale production, on which those services depend.

Whether the means of providing those services should be publicly or privately owned is a political question. In most industrial countries it is now generally felt that private ownership is to be preferred, but it is not generally perceived how monopolies in the private sector can be prevented from exploiting their customers' dependence on their services to increase their profits excessively at their customers' expense without stifling their inducement to economise.

Competition is widely assumed to be the best way of inducing privately owned companies to economise and pass on the benefits to their customers, and even a little competition is regarded as better than none. But competition has some severe disadvantages, even when there are many competitors, and it has proved to be very unsatisfactory when there are only a few.

Such considerations merit serious consideration of a new approach, described in Chapter Seven under the title 'economy-sharing regulation', whereby the permitted profits (or dividends) of a monopoly are tied by a formula to the economic benefit the monopoly gives its customers in the way of successive price reductions. The formula simulates the advantages of competition whilst avoiding its worst disadvantages. It also stimulates the monopoly to rationalise its prices and methods of charging as much as it can and to refrain from subsidising one set of customers at the expense of another. The formula is also applicable to oligopolies and could provide a framework within which the regulated companies could choose to compete or co-operate without external supervision. Thus competition and co-operation need no longer be regarded as mutually exclusive.

One political implication of these conclusions is that communism has failed and always will fail as a method of governing a society which aspires to prosper from technology, however benign its leaders may be. But that does not mean that the worst hardships imposed by unrestrained capitalism on the weaker parts of a society have to be accepted without question as the price of progress. A democratic society can prosper from technology without those hardships if some essential features of its organisation are derived from an understanding of technology.

The economic theory of rent needs to be revived to provide for the rational allocation of physically limited resources, but not for the enrichment of private owners of those resources. Unfortunately the term 'rent' has acquired emotive overtones of oppression from its historical association with aristocratic landlords living in idle affluence at the expense of their poor tenants, but that consequence can be avoided when the theory is revived.

Non-commercial public services, especially the National Health Service in Britain, present a serious – and as yet unsolved – problem. The cost method can provide any fully specified service at the lowest practicable cost if the method is understood and fully supported throughout the organisations which provide the service and the finance for it. But the absence of charges and prices for the service makes it impossible for the cost method to indicate what the service should comprise. Vague references to 'what the nation can afford' or 'best value for money' do not provide or lead to any answer. The problem is studiously ignored by most people.

Quasi-commercial services, notably subsidised public transport, present a similar problem, but in their case the solution (in the interest of prosperity from technology) is gradually to reduce the subsidies until the services are fully commercial. The almost universal objection to that policy can be seen to be mistaken if the emphasis on 'gradually' is kept firmly in place and the other features which have emerged from understanding technology (notably the revival of the theory of economic rent) are not overlooked.

The cost method provides no guidance on what constitutes 'fair' wages or other remuneration, nor does it support the notion that remuneration is or should be a reward for hard work as such. For a society to prosper from technology to the greatest possible extent, it must recognise that the purpose of remuneration is to persuade its free citizens to perform necessary tasks which would otherwise not be performed. Some jobs have to be paid more than others to attract enough people to do them – because they are tiring, dangerous, dirty, lonely or beyond the ability of most of the population.

If trade unions and/or professional associations succeed in raising their members' remuneration substantially above that level, or in forcing employers to employ more of their members than the jobs require (as they undoubtedly have done, at least in Britain), they thereby impoverish the rest of the community more than they enrich their members. The difficult – and as yet unsolved – problem is how to prevent them from doing so while protecting their weaker members from unrestrained exploitation by their employers.

Unemployment has to be accepted as part of the price of prosperity from technology. An understanding of the way in which technology progresses shows that central government must shoulder the full responsibility for training and retraining people as and when that progress puts them out of work. Technological considerations dictate that 'vocational' training, as it is called, is one of the few areas where full government participation and finance are imperative.

Although unemployment cannot be avoided, it happens that the cost method has a corollary, outlined in Chapter Eight, which has not been much debated but which would lessen the level of long-term unemployment in the worst-affected areas without increasing the flow of government money into those areas.

The second half of the twentieth century has seen the emergence of a set of enormous, previously unexpected threats to living standards everywhere. The threats arise from environmental pollution, which may be local, regional or global, and are typified by (but not restricted to) exhaust emissions from road vehicles, 'acid rain' mainly from electricity generating stations and global warming due to the increasing concentrations of carbon dioxide and other greenhouse gases in the atmosphere. Many people are already suffering from the pollution, but the threats are mainly of what will happen in the twenty-first and subsequent centuries if no redressing actions are taken, or even to some extent if they are.

An understanding of technology shows that, whether or not technology or technologists should be blamed for the pollution, technology offers the only way of tackling the problems.

Chapter Nine discussed some of the causes of pollution. It did not provide a packaged solution to any of them, but it offered a nine-point strategy for tackling them.

The three most important conclusions of that chapter are:

1) Pollution and pollution levels can be measured physically and sometimes predicted, but it is impossible to measure the harm they cause objectively. That can only be judged subjectively by or in consultation with the population which is or will be subjected to the pollution.
2) Technologists all over the world will have to strive as seldom before to tackle pollution and they will apply the cost method (except to the extent that they are persuaded or compelled not to) because it is the only rational method of applying technology.
3) It follows that their efforts should be directed by specifying the ends which are desired, leaving them the maximum scope for exercising their skills and ingenuity in achieving those ends at the lowest practicable cost.

On the bright side, tackling the problems of environmental pollution should lead to reductions in unemployment, unless the problems are very badly handled.

A society which aspires to prosper from technology will be disappointed if most of its population does not hold certain philosophical beliefs which are widely accepted in the Western world but much less firmly, if at all, in some other countries. They are the belief in freewill and personal responsibility for one's decisions and actions and in the importance of forethought.

The strength of those philosophical conclusions and the confidence with which they can be drawn have been increased by a historically recent development of academic philosophy. Before that development, the 'Holy Grail' of leading philosophers was a firm foundation for all rational thought, but now the search for such a foundation has been largely abandoned. Now it is recognised that rational thought must start from common sense, which is recognised as a vague starting point but from which progress can be made by demonstrating that some common sense beliefs are mistaken and should be replaced by new theories, which then gradually become part of common sense. That is how technologists think instinctively and it leads to those conclusions.

Many of the conclusions in this book have been reached by others on previous occasions, but as far as I am aware they have not previously been linked together and derived expressly from a knowledge and understanding of technology. I shall be content if that process has contributed to the resolution of some old controversies and provided some new insights into the features which societies must have in order to prosper from technology.